iPhone® Geekery

50 Insanely Cool Hacks and Mods for Your iPhone 4S

Guy Hart-Davis

New York Chicago San Francisco Lisbon
London Madrid Mexico City Milan New Delhi
San Juan Seoul Singapore Sydney Toronto

The McGraw·Hill Companies

Cataloging-in-Publication Data is on file with the Library of Congress

McGraw-Hill books are available at special quantity discounts to use as premiums and sales promotions, or for use in corporate training programs. To contact a representative, please e-mail us at bulksales@mcgraw-hill.com.

iPhone® Geekery: 50 Insanely Cool Hacks and Mods for Your iPhone 4S

1234567890 QFR QFR 1098765432

ISBN   978-0-07-179866-2
MHID   0-07-179866-8

Sponsoring Editor Roger Stewart	**Proofreader** Emily Rader	**Art Director, Cover** Jeff Weeks
Editorial Supervisor Janet Walden •	**Indexer** Jack Lewis	**Cover Designer** Ty Nowicki
Project Manager Anupriya Tyagi, Cenveo Publisher Services	**Production Supervisor** Jean Bodeaux	
Acquisitions Coordinator Molly Wyand	**Composition** Cenveo Publisher Services	
Copy Editor Bill McManus	**Illustration** Cenveo Publisher Services Lyssa Wald	

This book is dedicated to Teddy.

No iPhones were harmed during the writing of this book.

About the Author

Guy Hart-Davis is the author of more than 70 computer books, including *How to Do Everything: iPhone 4S*; *How to Do Everything: iPod touch*; *How to Do Everything: iPod & iTunes, Sixth Edition*; *The Healthy PC, Second Edition*; *PC QuickSteps, Second Edition*; *How to Do Everything with Microsoft Office Word 2007*; and *How to Do Everything with Microsoft Office Excel 2007*.

Contents

Acknowledgments

I'd like to thank the following people for their help with this book:

- Roger Stewart for proposing and developing the book
- Molly Wyand for handling the acquisitions end
- Bill McManus for editing the manuscript with a light touch
- Janet Walden for assisting with the production of the book
- Anupriya Tyagi for coordinating the production of the book
- Cenveo Publisher Services for laying out the pages
- Emily Rader for proofreading the book
- Jack Lewis for creating the index

Introduction

Do you want to take your iPhone to its limits—and then beyond them?

If so, this is the book for you.

This book shows you how to get the very most out of your iPhone by using to the max all the features Apple intends you to use—and then extend your iPhone with capabilities Apple *doesn't* intend it to have.

What Does This Book Cover?

Here's what this book covers:

- Chapter 1, "Music and Audio Geekery," kicks off by showing you how to sync music and other content onto your iPhone from multiple computers rather than a single computer. You then learn how to use your iPhone as your home stereo and car stereo, how to create custom ringtones for free from your songs, and how to share your music smoothly among all your devices. I show you how to record high-quality audio on your iPhone, play your guitar through your iPhone, record your live band on your iPhone, and even use your iPhone to replace your live band.

DOUBLE GEEKERY

Why Is This Book Better Than Other iPhone Books?

Unlike other iPhone books, this book assumes that you already know how to use your iPhone—how to make phone calls and FaceTime calls, browse the Web, install apps, and so on. (If you don't know how to do all of this, pick up a copy of my book *How to Do Everything: iPhone 4S*, also from McGraw-Hill, which covers this stuff in detail.)

iPhone Geekery assumes you're already an intermediate or advanced iPhone user—and that you want to become even more advanced. So it starts from that point, giving you a full book's worth of the good stuff you actually want, instead of grinding through all the basics you already know and then ending with a few pages of advanced material.

- Chapter 2, "Photo and Video Geekery," tells you how to put your videos and DVDs on your iPhone for viewing anywhere, either on the iPhone's screen or on a TV you connect it to. We then dig into photography: sharing your photos easily using Photo Stream, taking macro and panorama shots, and capturing high-quality self-portraits and time-lapse movies. Finally, we build a Steadicam rig to stabilize your iPhone when shooting video on the move, and then we look at how to view your computer's webcam on your iPhone from anywhere.

- Chapter 3, "iPhone as Your Main Computer Geekery," shows you how to tap into the iPhone's powerful computing capabilities and actually use your iPhone as your main computer. First, you learn pro tricks for entering text quickly and accurately with the onscreen keyboard, and then you see how to connect a Bluetooth keyboard so that you can hammer in text at full speed. We then go through creating full-fat business documents—Word, Excel, PowerPoint, and PDFs—on your iPhone and using your iPhone as not only a portable drive but also as a file server for your home network or workgroup. We finish the chapter by turning you into a power user of the vital Mail app and setting you up to give presentations directly from your iPhone.

- Chapter 4, "Security and Troubleshooting Geekery," teaches you how to secure your iPhone against theft or intrusion, how to track it down if you lose it, and how to wipe the data from your iPhone if you can't recover it. You also learn how to use your iPhone safely in wet or dirty conditions, troubleshoot software and hardware problems, and restore your iPhone to factory settings if it has software problems or if you're ready to sell it.

- Chapter 5, "Cellular, Wi-Fi, and Remote Geekery," starts by showing you how to unlock your iPhone from its carrier so that you can connect it to a different carrier's network. You then learn how to share your iPhone's Internet connection with your computers or devices, how to take control of your PC or Mac from your iPhone, and how to connect your iPhone to your company network across the Internet using a virtual private network. Finally, we discuss how to make calls using Voice over IP rather than the cellular network.

- Chapter 6, "Jailbreaking and Advanced Geekery," starts by showing you how to back up your iPhone for safety, and then shows you how to "jailbreak" your iPhone, freeing it from the restraints that Apple has placed upon it. You then discover how to find and install third-party apps that Apple hasn't approved, how to connect to your iPhone via SSH and transfer files to its file system, and how to recover any wasted space on the OS partition. You learn how to apply themes to change your iPhone's user interface, how to make Wi-Fi–only apps run over 3G connections, and how to play console and arcade games under emulation. And then we look at how to open your iPhone, put a custom back on it, and insert a near-field communications card for making contact-free payments. Finally—and only if you want to—we put your iPhone back in its Apple jail.

Conventions Used in This Book

To make its meaning clear without using far more words than necessary, this book uses a number of conventions, several of which are worth mentioning here:

- Double Geekery sidebars provide in-depth focus on important topics.

Note paragraphs highlight extra information you may find useful.

- The pipe character or vertical bar denotes choosing an item from a menu, usually on the PC or Mac but also sometimes on the iPhone. For example, "choose File | Open" means that you should click the File menu and select the Open item on it. Use the keyboard, mouse, or a combination of the two as you wish. Similarly, "choose Settings | General | About" means you tap the Settings icon on your iPhone's Home screen, tap the General button, and then tap the About button.

Tip paragraphs give you useful tricks, techniques, and workarounds.

- The ⌘ symbol represents the COMMAND key on the Mac—the key that bears the Apple symbol and the quad-infinity mark on most Mac keyboards.

Caution paragraphs help you steer around pitfalls.

- Most check boxes have two states: *selected* (with a check mark in them) and *cleared* (without a check mark in them). This book tells you to *select* a check box or *clear* a check box rather than "click to place a check mark in the box" or "click to remove the check mark from the box." Often, you'll be verifying the state of the check box, so it may already have the required setting—in which case, you don't need to click at all.

1 Music and Audio Geekery

Apple started its current line of portable devices with the iPod—so it's no surprise that your iPhone contains powerful features for playing music and working with audio.

At the most basic level, your iPhone is great for playing music on the go. I'm betting you know how to do this already, so we won't cover it in this book. But to put exactly the music you want on your iPhone, you will quite likely need to connect your iPhone to not just your main computer but also other computers. We'll start the chapter by looking at how to sync music and other content from multiple computers.

Next, we'll look at how you can use your iPhone as your home stereo and as your car stereo. We'll then explore how to create your own custom ringtones for free from your music and how to share your music among your iPhone, your computers, and your other iOS devices by using Apple's Home Sharing feature and the iCloud service.

Toward the end of the chapter, we'll examine how you can record high-quality audio on your iPhone, how you can play your guitar through your iPhone, how you can use your iPhone to record your live band, and how you can use your iPhone as your backing track for when your band is absent.

Project 1: Load Your iPhone with Content from Multiple Computers

As you know, you can sync your iPhone in either of two ways: by using Apple's iCloud online service, or by using your computer.

iCloud is arguably the wave of the future, and it's a great way to keep your music and other content synced among your iPhone, your computer, and other iOS devices you have (for example, an iPad or an iPod touch). In fact, you don't even need a computer—you can do it all with your iPhone or other iOS devices.

However, if you have a large collection of music and videos, or if you have a poky or unreliable Internet connection, you're probably better off syncing with your computer. Apple has got you covered here—you can download the latest version of iTunes, install it on your PC or Mac, and get your iPhone syncing underway inside a few minutes.

But what if you want to load content onto your iPhone from multiple computers rather than a single one?

Apple has designed your iPhone and iTunes to assume you'll be syncing all your information from a single computer. This is what many people—perhaps most—will do. But given that you're reading this book, you're most likely among those special ones who want to load your iPhone from multiple computers.

This project shows you how to do that. We'll start with the limitations.

Understand What You Can and Can't Sync

Here's what you need to know about syncing your iPhone with multiple computers:

- Your iPhone can sync only with a single iTunes library at a time. So if you sync your desktop's iTunes library with your iPhone, you can't then sync your laptop's iTunes library without wiping the desktop library from your iPhone.
- The iTunes library includes music, movies, TV shows, ringtones, podcasts, and books. To sync any of these items with the iTunes library of a computer other than your iPhone's home computer's iTunes library, you must wipe out your iPhone's existing library.
- Apps are separate from the iTunes library, but you can sync only one computer's set of apps to an iPhone. This can be a different computer than the computer whose iTunes library you're syncing for music, movies, and so on. Syncing apps with another computer removes all the existing apps from your iPhone. (Not the built-in apps—you'd need a virtual bulldozer to shift those.)
- Photos are also separate from the iTunes library you're using for syncing music, but you can sync photos from only one computer with your iPhone. Syncing photos with another computer removes your iPhone's existing photos (but not any photos or videos in your iPhone's Camera Roll).
- The items that appear on the Info tab of your iPhone's control screens—contact information, calendar information, mail accounts, bookmarks, and notes—are also handled separately from music. When you start to sync your iPhone's information items with another library, you can choose between merging the new information with the existing information and simply replacing the existing information.
- You can tell iTunes that you want to manage your iPhone's music and videos manually. After you tell iTunes this, you can connect your iPhone to a computer other than its home computer and add music and videos to it from that computer. But if you switch your home computer's library back to automatic syncing, you'll lose any music and videos you've added from other computers.
- For copyright reasons, iTunes puts limitations on which music and video files you can copy from your iPhone to a computer. For example, you can't connect your iPhone to your friend's computer and copy all the songs from your iPhone to the computer. See the sidebar at the end of this project for instructions on working around this limitation—for example, to recover your iTunes library after your computer crashes.

Set Your iPhone to Sync Data from Multiple Computers

Now that you understand the restrictions on syncing, let's look at how to set your iPhone to sync data from multiple computers rather than a single computer.

 When you set your iPhone to sync music or photos with another computer, the initial sync may take hours, because of the amount of data involved. By contrast, syncing information (contacts, calendars, and so on) usually takes only seconds, and syncing apps takes a few minutes, depending on how many apps there are and how chunky their developers have made them.

Sync All Your iPhone's Data with Its Current Computer

Before you start making changes, connect your iPhone to its current home computer and run a sync. This makes sure that you have a copy of the latest information from your iPhone on your computer in case you need it later.

Change the iTunes Library Your iPhone Is Syncing Music with

To change the iTunes library your iPhone is syncing music with, follow these steps:

1. Connect your iPhone to the computer that contains the music you want to sync.
2. Click your iPhone's entry in the Devices category of the Source list in the iTunes window to display its control screens.
3. Click the Music tab to display the Music screen.
4. Select the Sync Music check box.
5. Use the controls to specify which music you want to sync (see Figure 1-1). For example, either select the Entire Music Library option button to sync the whole library (assuming your iPhone has enough room for it), or select the Selected Playlists, Artists, Albums, And Genres option button, and then select the check box for each item you want to include.

 At this point, you can also choose sync settings for the other items that changing the music library will affect: ringtones, movies, TV shows, podcasts, and books.

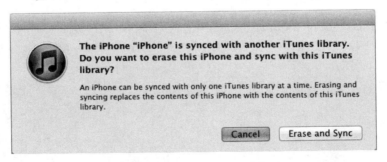

FIGURE 1-1 On the Music screen, choose whether to sync all the music in your library or only the playlists, artists, albums, and genres whose check boxes you select.

6. Click the Apply button, which replaces the Sync button when you make changes. iTunes displays a dialog box asking if you want to erase and sync the library, as shown here.

The iPhone "iPhone" is synced with another iTunes library. Do you want to erase this iPhone and sync with this iTunes library?

An iPhone can be synced with only one iTunes library at a time. Erasing and syncing replaces the contents of this iPhone with the contents of this iTunes library.

Cancel   Erase and Sync

7. Click the Erase And Sync button. iTunes replaces the existing library items on your iPhone with the items you chose from the new library.

Change the Computer Your iPhone Is Syncing Information With

To change the computer your iPhone is syncing contacts, calendars, mail accounts, and other information with, follow these steps:

1. Connect your iPhone to the computer that contains the information you want to sync.
2. Click your iPhone's entry in the Devices category in the Source list in the iTunes window to display its control screens.
3. Click the Info tab to display the Info screen (see Figure 1-2).

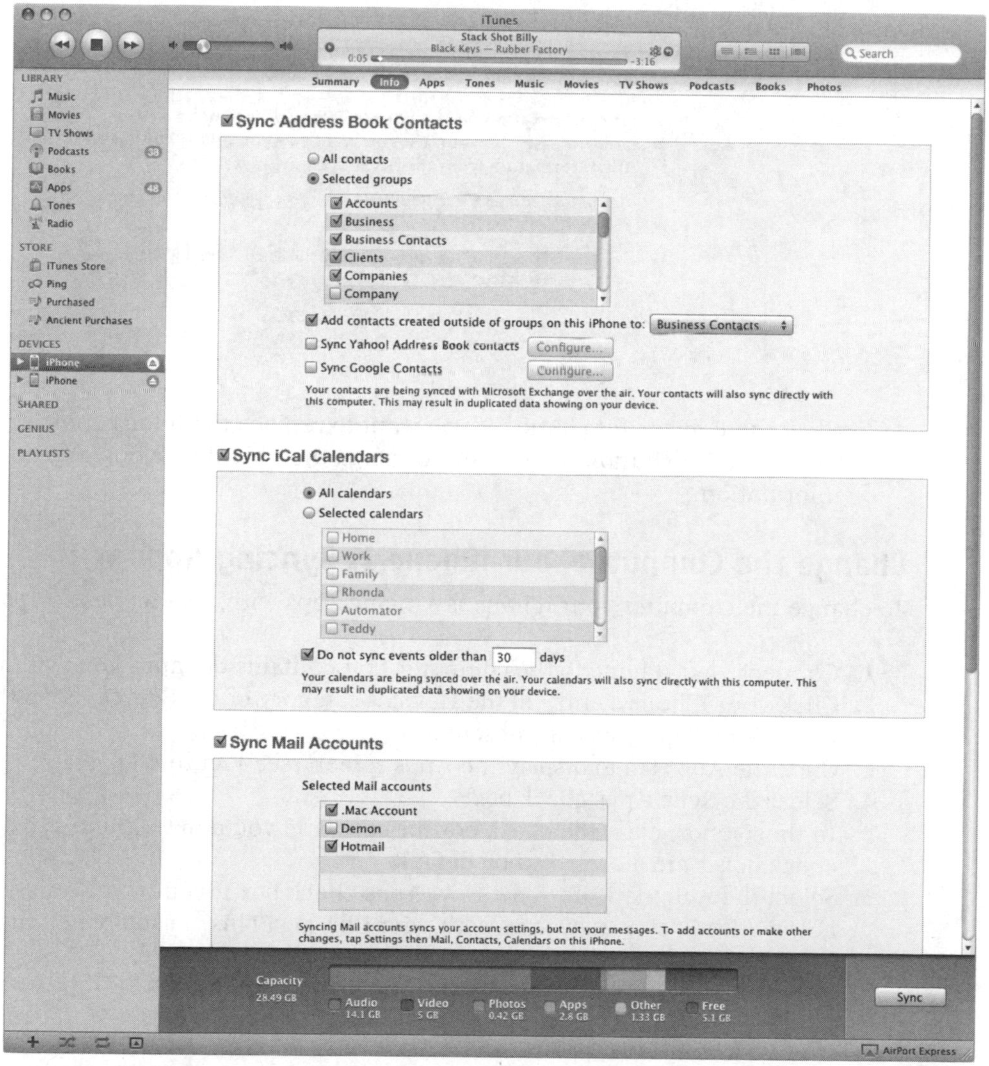

FIGURE 1-2   On the Info screen, choose which contacts, calendars, mail accounts, bookmarks, and notes to sync from your computer to your iPhone.

4. Select the appropriate check boxes. For example, on a Mac, select the Sync Address Book Contacts check box, the Sync iCal Calendars check box, and the Sync Mail Accounts check box; and select the Sync Safari Bookmarks check box and the Sync Notes check box in the Other box as needed.

5. Use the controls in each box to specify which items you want to sync. For example, select the check box for each mail account to sync in the Selected Mail Accounts list box.

6. Click the Apply button, which replaces the Sync button when you make changes. iTunes displays a dialog box (shown here) that offers you the choice between replacing the information on your iPhone and merging the new information with the existing information.

The information on the iPhone "iPhone" is synced with another user account. Do you want to sync this iPhone with the information from this user account instead?

"Merge Info" merges the information on this iPhone with the information from this user account.

"Replace Info" replaces the information on this iPhone with the information from this user account.

| Cancel | | Replace Info | Merge Info |

7. Click the Replace Info button if you want to replace the information, or click the Merge Info button if you want to merge the old information and new information.

Change the Computer Your iPhone Is Syncing Apps With

To change the computer your iPhone is syncing apps with, follow these steps:

1. Connect your iPhone to the computer that contains the apps you want to sync.

2. Click your iPhone's entry in the Devices category of the Source list in the iTunes window to display its control screens.

3. Click the Apps tab to display the Apps screen (see Figure 1-3).

4. Select the Sync Apps check box.

5. In the list box, clear the check box for each app you don't want to sync. These check boxes are all selected by default.

6. Select the Automatically Sync New Apps check box if you want iTunes to automatically sync new apps with your iPhone. (This is usually helpful.)

FIGURE 1-3   On the Apps screen, choose which apps to sync with your iPhone.

7. Click the Apply button. iTunes displays a dialog box (shown here) to confirm that you want to replace all your iPhone's apps with the apps in this computer's iTunes library.

8. Click the Sync Apps button. iTunes syncs the apps.

Change the Computer Your iPhone Is Syncing Photos With

To change the computer your iPhone is syncing photos with, follow these steps:

1. Connect your iPhone to the computer that contains the photos you want to sync.
2. Click your iPhone's entry in the Devices category of the Source list in the iTunes window to display its control screens.
3. Click the Photos tab to display the Photos screen (see Figure 1-4).
4. Select the Sync Photos From check box.
5. In the Sync Photos From drop-down list, choose the source of the photos. For example, choose the Pictures folder on Windows or choose iPhoto on Mac OS X.
6. Use the controls to specify which photos to sync. For example, select the All Photos, Albums, Events, And Faces option button if you want to sync all the photos (assuming they'll fit on your iPhone). Or select the Selected Albums, Events, And Faces, And Automatically Include option button, choose a suitable item in the drop-down list, and then select the check box for each album, event, and face you want to sync.

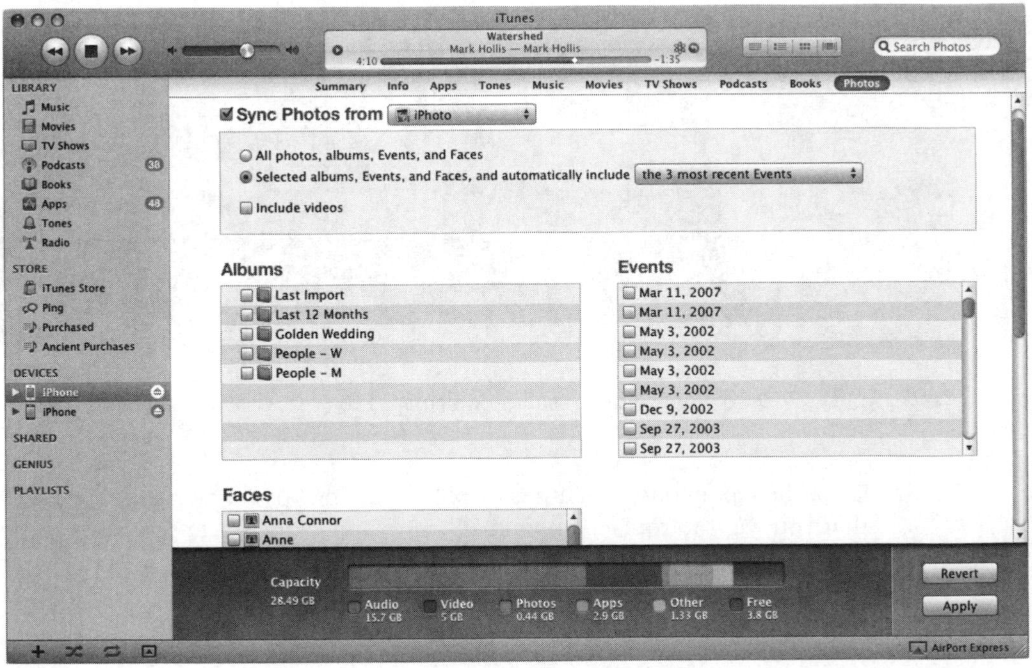

FIGURE 1-4   On the Photos screen, choose which photos to sync with your iPhone.

7. Click the Apply button. iTunes displays a dialog box (shown here) to confirm that you want to replace the synced photos on your iPhone.

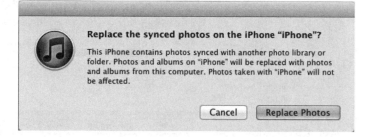

Replace the synced photos on the iPhone "iPhone"?

This iPhone contains photos synced with another photo library or folder. Photos and albums on "iPhone" will be replaced with photos and albums from this computer. Photos taken with "iPhone" will not be affected.

Cancel   Replace Photos

8. Click the Replace Photos button. iTunes replaces the photos.

 If you're using a Mac, you can use iPhoto or Image Capture to copy photos from your iPhone to your Mac. Use iPhoto when you want to gather your photos into Events, edit them, and manage them in iPhoto. Use Image Capture when you just want to get the photos (or screen captures, or saved images) from your iPhone into your Mac's file system.

 DOUBLE GEEKERY

Recover Your Songs and Videos from Your iPhone

When you sync your iPhone with your computer, any songs and videos on your iPhone are also in your library on your computer, so you don't need to transfer the songs and videos from your iPhone to your computer. This includes the songs and videos you purchase on your iPhone from the iTunes Store. But if you have a computer disaster, or if your computer is stolen, you may need to recover the songs and videos from your iPhone to your new or repaired computer.

To recover songs and videos from your iPhone, you need a utility that can read your iPhone's file system. To help you avoid losing your music and videos, iPhone enthusiasts have developed some great utilities for transferring files from your iPhone's hidden music and video storage to a computer.

At this writing, several utilities are available for copying your music and videos from your iPhone to your computer. The best utility for both Windows and the Mac is DiskAid from DigiDNA ($24.90; www.digidna.net; trial version available). DiskAid (shown next) reads your iPhone's library database and displays its contents so that you can easily copy them back to a computer.

Project 2: Use Your iPhone as Your Home Stereo

Your iPhone is great for music on the go, but you can use it as your home stereo as well. In this project, we'll look at the five best ways of doing so:

- Using an iPhone speaker dock
- Connecting your iPhone directly to your stereo with a cable
- Connecting your iPhone to your stereo or speakers via Bluetooth or a radio transmitter
- Using the AirPlay feature to play music to an AirPort Express
- Quickly boosting the bass with an echo chamber when you're out and about

Use an iPhone Speaker Dock

The simplest way to get a decent volume of sound from your iPhone is to connect it to a pair of powered speakers (speakers that include their own amplifier). You can buy speakers designed especially for the iPhone, which use the Dock Connector port for high-quality output. But you can also get good sound using your iPhone with any powered speakers that accept input via a miniplug connector (the size of connector used for the iPhone's headphones).

DOUBLE GEEKERY

Why You Should Connect Speakers to the Dock Connector Port Rather Than the Headphone Port

When you connect external speakers to your iPhone, you have a choice of ports: the headphone port or the Dock Connector port.

If possible, use the Dock Connector port rather than the headphone port. You can use it either directly with a speaker or cable that has a Dock Connector, or indirectly by connecting your iPhone to a dock and then connecting the speakers to the line-out port on the dock.

The Dock Connector port delivers a fixed output level and better audio quality than the headphone port (whose output level varies depending on the volume setting), so it's a much better choice. Most speakers designed specifically for use with the iPhone have a Dock Connector that enables them to receive audio at line-out quality and a constant volume.

When you need to connect speakers to the headphone port rather than the Dock Connector port, turn your iPhone's volume all the way down at first. The headphone port puts out up to 60 milliwatts (mW) altogether—30 mW per channel—and can deliver a high enough signal to cause distortion or damage to an input that's expecting a standard line-out volume. After you make the connection, start playing audio and turn your iPhone's volume up gradually until you get a suitable level on the input.

Connect Your iPhone to Your Existing Stereo

If you have a good stereo, you can play music from your iPhone through it. In this section, we'll look at how to connect your iPhone to your stereo using a cable, using Bluetooth, and using a radio transmitter.

Connect Your iPhone to a Stereo with a Cable

The most direct way to connect your iPhone to a stereo system is with a cable. For a typical receiver, you'll need a cable that has a miniplug at one end and two RCA plugs at the other end. Figure 1-5 shows an example of an iPhone connected to a stereo via the amplifier.

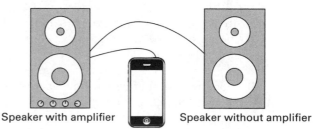

Speaker with amplifier   Speaker without amplifier

FIGURE 1-5   A miniplug-to-RCA-plugs cable is the most direct way of connecting an iPhone to your stereo system.

Some receivers and boom boxes use a single stereo miniplug input rather than two RCA ports. To connect your iPhone to such devices, you'll need a stereo miniplug-to-miniplug cable. Make sure the cable is stereo, because mono miniplug-to-miniplug cables are common. A stereo cable has two bands around the miniplug (as on most headphones), whereas a mono cable has only one band.

If you have a high-quality receiver and speakers, get a high-quality cable to connect your iPhone to them. After the amount you've presumably spent on your iPhone and stereo, it'd be a mistake to degrade the signal between them by sparing a few bucks on the cable.

 You can find various home-audio connection kits that contain a variety of cables likely to cover your needs. These kits are usually a safe buy, but unless your needs are peculiar, you'll end up with one or more cables you don't need. So if you do know which cables you need, make sure a kit offers a cost savings before buying it instead of the individual cables.

Connect your iPhone to your receiver as follows:

1. Connect the miniplug to your iPhone's headphone port. If you have a dock, connect the miniplug to the dock's line-out port instead, because this gives a more consistent volume and better sound quality than the headphone port.
2. If you're using the headphone port, turn down the volume on your iPhone all the way.
3. Whichever port you're using, turn down the volume on the amplifier as well.
4. Connect the RCA plugs to the left and right ports of one of the inputs on your amplifier or boom box—for example, the AUX input or the Cassette input (if you're not using a cassette deck).

 Don't connect your iPhone to the Phono input on your amplifier. The Phono input is built with a higher sensitivity to make up for the weak output of a record player. Putting a full-strength signal into the Phono input will probably blow it.

5. Start the music playing. If you're using the headphone port, turn up the volume a little.
6. Turn up the volume on the receiver so that you can hear the music.
7. Increase the volume on the two controls in tandem until you reach a satisfactory sound level.

 Too low a level of output from your iPhone may produce noise as your amplifier boosts the signal. Too high a level of output from your iPhone may cause distortion.

Use a Bluetooth Connection Between Your iPhone and Your Stereo

Connecting your iPhone to your stereo with a cable gives you great quality, but it means your iPhone is anchored in place (unless you get a cable long enough for you to roam the room like a 1970s guitarist). If you want to keep your iPhone at hand while you play music through your stereo, try Bluetooth instead.

To connect your iPhone to your stereo via Bluetooth, get a device such as the Belkin Bluetooth Music Receiver ($49.95; http://store.apple.com and other retailers). This is a Bluetooth device that connects via a cable to your stereo. You then connect your iPhone to the receiver via Bluetooth and play music across the airwaves. Quality is lower than with a cable, but you'll probably feel the added freedom makes up for it.

Use a Radio Transmitter Between Your iPhone and a Stereo

If you don't want to connect your iPhone directly to your stereo system or pay for a Bluetooth receiver, you can use a radio transmitter to send the audio from your iPhone to the radio on your stereo.

The sound you get from this arrangement typically will be lower in quality than the sound from a wired connection, but it should be at least as good as listening to a conventional radio station in stereo. If that's good enough for you, a radio transmitter can be a neat solution to playing music from your iPhone throughout your house.

 Using a radio transmitter has another advantage: You can play the music on several radios at the same time, giving yourself music throughout your dwelling without complex and expensive rewiring.

To use the radio transmitter, you connect it to your iPhone's headphone socket or Dock Connector port, set the frequency to transmit, and then set the music playing. When you tune your radio in to that frequency, it receives the broadcast just like a regular radio station.

For more detail on using a radio transmitter, see the section "Use a Radio Transmitter with Your iPhone" in the next project.

Use Your iPhone's AirPlay Feature to the Max

Your iPhone includes a feature called AirPlay that enables it to play music on remote speakers connected to an AirPort Express wireless access point.

Play Audio to an AirPort Express

If you have an AirPort Express (a wireless access point that Apple makes), you can use it not only to network your home but also to play music from your iPhone or your computer through your stereo system.

To play music through an AirPort Express, first set it up like this:

1. Connect the AirPort Express to the receiver via a cable. The line-out port on the AirPort Express combines an analog port and an optical output, so you can connect the AirPort Express to the receiver in either of two ways:
   - Connect an optical cable to the AirPort Express's line-out socket and to an optical digital-audio input port on the receiver. If the receiver has an optical input, use this arrangement to get the best sound quality possible.
   - Connect an analog audio cable to the AirPort Express's line-out socket and to the RCA ports on your receiver.
2. If your network has a wired portion, connect the Ethernet port on the AirPort Express to the switch or hub using an Ethernet cable. If you have a DSL that you will share through the AirPort Express, connect the DSL via the Ethernet cable.
3. Plug the AirPort Express into an electric socket.

You can now play music from your iPhone by tapping the AirPlay icon, the icon showing a solid triangle superimposed on a hollow rectangle, as shown in the lower-right corner of the left screen in Figure 1-6. In the AirPlay dialog box that opens (shown on the right in Figure 1-6), tap the AirPort Express button.

When you need to switch back to your iPhone's speakers, tap the AirPlay icon again, and this time tap your iPhone button.

FIGURE 1-6 Tap the AirPlay icon (the triangle and rectangle in the lower-right corner of the left screen) to display the AirPlay dialog box (right), and then tap the AirPort Express button.

To play music from iTunes on your computer through the AirPort Express, you use a similar technique. Click the AirPlay icon near the lower-right corner of the iTunes window to display the pop-up menu with the speakers you can use, as shown here. Then click the AirPort Express item on the menu to direct iTunes' output to the AirPort Express.

When playing from iTunes, you can also play music through both an AirPort Express and through your computer's own speakers. To do this, follow these steps:

1. In iTunes, click the AirPlay icon to display the pop-up menu.
2. Click the Multiple Speakers item to display the Multiple Speakers dialog box, shown here.

3. Select the check box for each speaker set you want to use.
4. Drag the volume sliders to adjust the relative volume of the speakers.
5. Click the Close button (the × button on Windows, the red button in the upper-left corner on the Mac) to close the Multiple Speakers dialog box.

Boost the Bass with a Pint Glass

When you're out and about with your iPhone, you probably don't carry a pair of speakers with you. If you're somewhere where you can patch your iPhone into a stereo or use AirPlay to play into an AirPort Express, you're set. But otherwise you need to rely on your iPhone's speaker, which doesn't have the muscle for playing to a group of any size.

Here's a quick remedy: Grab a pint glass and make sure it's empty and (preferably) clean and dry. Set your music playing, and then place your iPhone in the glass, as shown in Figure 1-7. The glass acts as an echo chamber, giving you about double the volume, and considerably more bass. Quality is pretty good, considering what you're using.

When you finish playing music with the pint glass, use it for its normal purpose.

FIGURE 1-7 Use a pint glass as an echo chamber to boost your iPhone's output.

Project 3: Use Your iPhone as Your Car Stereo

You can connect your iPhone to a car stereo in any of the following ways:

- Get a car with a built-in iPhone connection or add an after-market iPhone-integration device.
- Use a cassette adapter to connect your iPhone to the car's cassette player.
- Use a radio transmitter to play your iPhone's output through the car's radio.
- Wire your iPhone directly to the car stereo and use it as an auxiliary input device.

Each of these methods has its pros and cons. The following sections tell you what you need to know to choose the best option for your car stereo.

Use a Built-in iPhone Connection

At this writing, Apple claims that more than 90 percent of new cars sold in the United States have an option for connecting an iPhone or iPod. (See the list at www.apple.com/ipod/car-integration.) So if you're in the market for a new car, add iPhone connectivity to your list of criteria. Similarly, if you're buying a used car that's only a few years old, you may be able to get iPhone connectivity built in.

If your car doesn't have its own means of integrating an iPhone, look for a third-party solution. You'll find a list of third-party solutions at the bottom of the www.apple.com/ipod/car-integration web page. Some adapters not only let you play back music from your iPhone through the car's stereo and control it using the stereo system's controls, but also let you display the song information from your iPhone on the stereo's display, making it easier to see what you're listening to.

Use a Cassette Adapter with Your iPhone

If the car stereo has a cassette player, your easiest option is to use a cassette adapter to play audio from your iPhone through the cassette deck. You can buy such adapters for between $10 and $20 from most electronics stores or from an iPhone specialist.

The adapter is shaped like a cassette and uses a playback head to input analog audio via the head that normally reads the tape as it passes. A wire runs from the adapter to your iPhone.

A cassette adapter can be an easy and inexpensive solution, but it's far from perfect. The main problem is that the audio quality tends to be poor, because the means of transferring the audio to the cassette player's mechanism is less than optimal. But if your car is noisy, you may find that road noise obscures most of the defects in audio quality.

If the cassette player's playback head is dirty from playing cassettes, audio quality will be that much worse. To keep the audio quality as high as possible, clean the cassette player regularly using a cleaning cassette.

If you use a cassette adapter in an extreme climate, try to make sure you don't bake it or freeze it by leaving it in the car.

Use a Radio Transmitter with Your iPhone

If the car stereo doesn't have a cassette deck, your easiest option for playing music from your iPhone may be to get a radio transmitter. This device plugs into your iPhone and broadcasts a signal on an FM frequency to which you then tune your radio to play the music. Better radio transmitters offer a choice of frequencies to allow you easy access to both your iPhone and your favorite radio stations.

Radio transmitters can deliver reasonable audio quality. If possible, try before you buy by asking for a demonstration in the store (take a portable radio with you, if necessary).

The main advantages of these devices are that they're relatively inexpensive (usually between $15 and $50) and they're easy to use. They also have the advantage that you can put your iPhone out of sight (for example, in the glove compartment—provided it's not too hot) without any telltale wires to help the light-fingered locate it.

On the downside, most of these devices need batteries (others can run off the 12-volt accessory outlet or cigarette-lighter socket), and less expensive units tend not to deliver the highest sound quality. The range of these devices is minimal, but at close quarters, other radios nearby may be able to pick up the signal—which could be embarrassing, entertaining, or irrelevant, depending on the circumstances. If you use the radio transmitter in an area where the airwaves are busy, or you drive through areas that use different frequencies, you may need to keep switching the frequency to avoid having the transmitter swamped by the full-strength radio stations.

 DOUBLE GEEKERY

Find a Suitable Frequency for a Radio Transmitter

In most areas, the airwaves are busy these days—so to get good reception on your car's radio from your iPhone's radio transmitter, you need to pick a suitable frequency. To do so, follow these steps:

1. With your iPhone's radio transmitter turned off, turn on your car radio.
2. Tune the car radio to a frequency on which you get only static, and for which the frequencies one step up and one step down give only static as well. For example, if you're thinking of using the 91.3 frequency, make sure that 91.1 and 91.5 give only static as well.
3. Tune the radio transmitter to the frequency you've chosen, and see if it works. If not, identify and test another frequency.

This method may sound obvious, but what many people do is pick a frequency on the radio transmitter, tune the radio to it—and then be disappointed by the results.

If you decide to get a radio transmitter, you'll need to choose between getting a model designed specifically for the iPhone and getting one that works with any audio source. Radio transmitters designed for the iPhone typically mount on the iPhone, making them a neater solution than general-purpose ones that dangle from the headphone socket. Radio transmitters designed for use with iPhones in cars often mount on the accessory outlet or dash and secure the device as well as transmitting its sound.

 A radio transmitter works with radios other than car radios, so you can use one to play music through your stereo system (or someone else's). You may also want to connect a radio transmitter to a PC or Mac and use it to broadcast audio to a portable radio. This is a great way of getting streaming radio from the Internet to play on a conventional radio.

Wire Your iPhone Directly to a Car Stereo

If neither the cassette adapter nor the radio transmitter provides a suitable solution, or if you simply want the best audio quality you can get, connect your iPhone directly to your car stereo. How easily you can do this depends on how the stereo is designed:

- If your car stereo has a miniplug input built in, get a Dock Connector–to-miniplug cable to connect your iPhone's Dock Connector port to the miniplug input. You can also use a miniplug-to-miniplug cable from your iPhone's headphone port to the miniplug input, but the Dock Connector gives you better quality.
- If your stereo is built to take multiple inputs—for example, a CD player (or changer) and an auxiliary input—you may be able to simply run a wire from unused existing connectors. Then all you need to do is plug your iPhone into the other end and press the correct buttons to get the music going.
- If no unused connectors are available, you or your local friendly electronics technician may need to get busy with a soldering iron.

If you're buying a new car stereo, look for iPhone integration or at least an auxiliary input that you can use with your iPhone.

Project 4: Use Home Sharing and Library Sharing to the Max

Both iTunes and your iPhone are designed to share your music with other computers and devices. You can share music either among computers and devices linked to the same iTunes account or with any compatible computer or device on the same network.

Understand the Difference Between Home Sharing and Library Sharing

iTunes gives you two different types of sharing:

- **Home Sharing**   Home Sharing lets you share your entire library with up to five computers and with your iPhone, iPod touch, or iPad. You can copy media files from one computer to another, so you can make sure each of your computers contains the same library. You can also set Home Sharing to automatically copy any new media files or apps you buy.
- **Library Sharing**   Library Sharing lets you share either your entire library or selected playlists with other computers on your network. The other computers can only play the songs or other media files; they cannot copy the files.

 The big difference between Home Sharing and iTunes' Library Sharing is that Home Sharing enables you to copy files, whereas Library Sharing doesn't. Home Sharing is for sharing media files among your computers; Library Sharing is for sharing your media files with other people.

To use Home Sharing, you set up each of the computers to use the same Apple ID. Using the same Apple ID is the mechanism for making sure that you're not violating copyright by giving copyrighted content to other people. If you don't have an Apple ID yet, you can create one from the Home Sharing screen.

 Your iPhone can access both the libraries you share via Home Sharing and libraries or playlists you share via Library Sharing. Your iPhone can also access libraries or playlists other people share via Library Sharing.

Set Up Home Sharing on Each Computer

To set up Home Sharing, follow these steps:

1. Open iTunes.
2. In the Source list on the left, see if the Shared category is expanded, showing its contents. If not, expand it by holding the mouse pointer over the Shared heading and then clicking the word Show when it appears.
3. Click the Home Sharing item to display its contents.
4. Type your Apple ID in the Apple ID box.

 If you don't yet have an Apple ID, click the Need An Apple ID? link, and then follow through the process of signing up for one. Once you're armed with your Apple ID, go back to the Home Sharing screen.

5. Type your password in the Password box.

6. Click the Create Home Share button. iTunes checks in with the iTunes servers and sets up the account.

 If iTunes displays a dialog box saying that Home Sharing could not be activated because this computer is not authorized for the iTunes account associated with the Apple ID you provided, click the Authorize button.

7. When the Home Sharing screen displays the message that Home Sharing is now on, click the Done button. iTunes then removes the Home Sharing item from the Shared category in the Source list, and you have access to the libraries of the other computers on which you've set up Home Sharing.

Copy Files Using Home Sharing

After setting up Home Sharing, you can quickly copy files from one installation of iTunes to another. To do so, follow these steps:

1. In the Source list in iTunes, make sure the Shared category is expanded, showing its contents. If the Shared category is collapsed, expand it by holding the mouse pointer over the Shared heading and then clicking the word Show when it appears.
2. Click the Home Sharing library whose contents you want to see. The library's contents appear in the main part of the iTunes window, and you can browse them as usual (see Figure 1-8). For example, choose View | Column Browser | Show Column Browser to display the column browser so that you can browse by genres, artists, albums, or whichever other items you prefer.

 The Home Sharing libraries appear in the Shared category with a Home Sharing icon next to them. The Home Sharing icon shows a house containing a musical note.

3. In the Show drop-down list at the bottom of the iTunes window, choose which items to display:
   • **All Items**   This is the default setting. Use it when you want to get an overview of what the library contains.
   • **Items Not In My Library**   Use this setting to display only the items you may want to copy to your library.
4. Select the items you want to import to your library. If you've switched to the Items Not In My Library view, you may want to choose Edit | Select All (or press CTRL-A on Windows or ⌘-A on the Mac) to select everything.
5. Click the Import button. iTunes imports the files.

FIGURE 1-8   You can browse a Home Sharing library using the same techniques as for browsing your own library.

DOUBLE GEEKERY

Make Home Sharing Automatically Import New Purchases from Your Other Computers

You can set Home Sharing to automatically import your new purchases from the iTunes Store to your computer. So if you buy a song on your laptop computer, you can have iTunes automatically import it to your desktop computer as well. If you buy a song on your iPhone from the iTunes Store, iTunes syncs the song first to whichever computer you use for syncing, and then imports it to the other computers you've set up for Home Sharing.

To set Home Sharing to automatically import new purchases, follow these steps:

1. In the Source list in iTunes, click a Home Sharing library to display its contents and the Home Sharing control bar.

2. Click the Settings button to display the Home Sharing Settings dialog box (shown here).

3. Select the Music check box, the Movies check box, the TV Shows check box, the Books check box, and the Apps check box, as needed.

4. Click the OK button to close the Home Sharing Settings dialog box.

Set Up Home Sharing on Your iPhone

Next, you need to set up Home Sharing on your iPhone to enable it to access the libraries you've shared using Home Sharing in iTunes.

To set up Home Sharing on your iPhone, follow these steps:

1. Press the Home button to display the Home screen.
2. Tap the Settings icon to display the Settings screen.
3. Scroll down to the third box, the one that starts with the General button (shown on the left in Figure 1-9).
4. Tap the Music button to display the Music screen (shown on the right in Figure 1-9).
5. In the Home Sharing box at the bottom, tap the Apple ID box, and then type your Apple ID and password.
6. Tap the Settings button to go back to the Settings screen.

FIGURE 1-9   On the Settings screen (left), tap the Music button to display the Music screen (right), and then enter your Apple ID and password in the Home Sharing box.

You can now access your shared libraries from the Music app. See the section "Play Shared Music from Your iPhone," later in this chapter, for details.

Set Up Library Sharing in iTunes on Your Computer

You can share either your entire library or selected playlists with other users on your network. You can share most items, including MP3 files, AAC files, Apple Lossless Encoding files, AIFF files, WAV files, and links to radio stations. You can't share Audible files or QuickTime sound files.

 Technically, iTunes' sharing is limited to computers on the same TCP/IP subnet as your computer is on. (A *subnet* is a logical division of a network.) A home network typically uses a single subnet, so your computer can "see" all the other computers on the network. But if your computer connects to a medium-sized network, and you're unable to find a computer that you know is connected to the same network somewhere, it may be on a different subnet.

At this writing, you can share your library with up to five other computers per day, and your computer can be one of up to five computers accessing the shared library on another computer on any given day.

The shared library remains on the computer that's sharing it, and when a participating computer goes to play a song or other item, that item is streamed across the network. This means that the item isn't copied from the computer that's sharing it to the computer that's playing it in a way that leaves a usable file on the playing computer.

When a computer goes offline or is shut down, library items it has been sharing stop being available to other users. Participating computers can play the shared items but can't do anything else with them; for example, they can't burn shared songs to CD or DVD, download them to an iPod or iPhone, or copy them to their own libraries.

To share either your entire library or selected playlists with other users of iTunes or iPhones (or iPod touches, or iPads) on your network, follow these steps:

1. Display the iTunes dialog box or the Preferences dialog box:
   - In Windows, choose Edit | Preferences or press CTRL-COMMA or CTRL-Y to display the iTunes dialog box.
   - On the Mac, choose iTunes | Preferences or press ⌘-COMMA or ⌘-Y to display the Preferences dialog box.
2. Click the Sharing tab to display it. Figure 1-10 shows the Sharing tab of the iTunes dialog box with settings chosen.
3. Select the Share My Library On My Local Network check box. (This check box is cleared by default.) By default, iTunes then selects the Share Entire Library option button. If you want to share only some playlists, select the Share Selected Playlists option button. Then, in the list box, select the check box for each playlist you want to share.

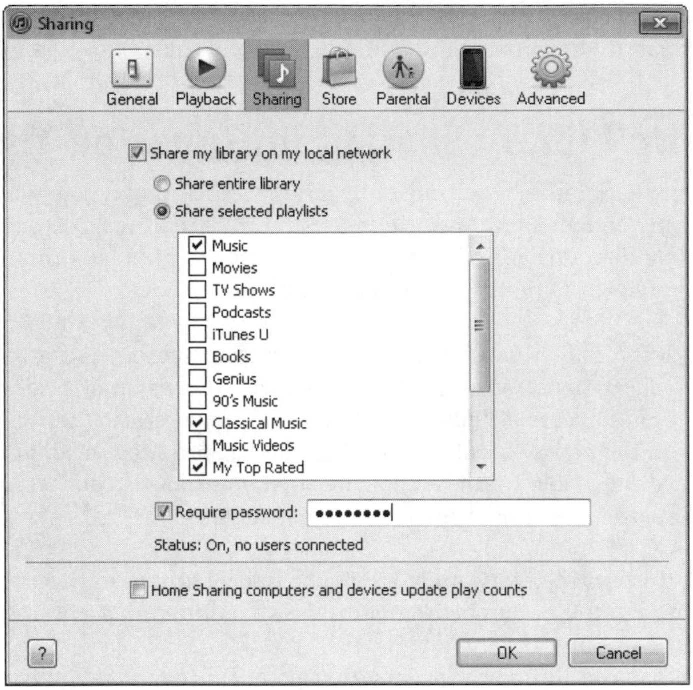

FIGURE 1-10   On the Sharing tab of the iTunes dialog box or the Preferences dialog box, choose whether to share part or all of your library.

4. By default, your shared library items are available to any other user on the network. To restrict access to people with whom you share a password, select the Require Password check box, and then enter a strong (unguessable) password in the text box.

> If there are many computers on your network, use a password on your shared music to help avoid running up against the five-users-per-day limit. If your network has only a few computers, you may not need a password to avoid reaching this limit.

5. Select the Home Sharing Computers And Devices Update Play Counts check box if you want iTunes to update the play count for a song whenever any computer plays it, not just this computer.
6. Click the General tab to display its contents. In the Library Name text box near the top of the dialog box, set the name that other users trying to access your library will see. The default name is *username*'s Library, where *username* is your username—for example, Anna Connor's Library. You might choose to enter a more descriptive name, especially if your computer is part of a well-populated network (for example, in a dorm).
7. Click the OK button to apply your choices and close the dialog box.

When you set iTunes to share your library, iTunes displays a message reminding you that "Sharing music is for personal use only"—in other words, remember not to violate copyright law. Select the Do Not Show This Message Again check box if you want to prevent this message from appearing again.

Play Shared Music from Your iPhone

To play shared music from your iPhone, follow these steps:

1. Press the Home button to display the Home screen.
2. Tap the Music button to display the Music app.
3. Tap the More button to display the More screen (shown on the left in Figure 1-11).
4. Tap the Shared button to display the Shared screen (shown on the right in Figure 1-11).
5. Tap the shared music library you want to access.

 If the Shared button doesn't appear on the More screen, Home Sharing is turned off on your iPhone. Turn it on as described in the section "Set Up Home Sharing on Your iPhone," earlier in this chapter.

FIGURE 1-11   Tap the Shared button on the More screen (left) to display the Shared screen (right), and then tap the shared music library you want to access.

Project 5: Create Your Own Custom Ringtones for Free

To make your iPhone sound unique and to give yourself a clear indication of when you receive phone calls, texts, voicemail, tweets, and so on, you can create custom ringtones and sync them to your iPhone. This is a great way to get ringtones that you not only like but that enable you to distinguish crucial calls from ignorable calls without having to look at your iPhone's screen to see the caller's name.

 Earlier versions of iTunes included a feature for making ringtones from songs bought from the iTunes Store. But Apple has removed this feature from iTunes 10, so you need to create your ringtones manually as described here.

To create a ringtone from a song, follow these steps:

1. Play the song and identify the part you want to use. This can be up to 30 seconds long. Note down the start time and end time.
2. Right-click (or CTRL-click on the Mac) the song, and then click Get Info on the context menu to display the Item Information dialog box for the song.

 The Item Information dialog box doesn't actually show the words "Item Information" in its title bar. On Windows, the title bar shows the word "iTunes"; on the Mac, the title bar shows the song's title rather than the words "Item Information."

3. Click the Options tab to bring it to the front of the Item Information dialog box (see Figure 1-12).
4. Click in the Start Time box and enter the start time for the ringtone section—for example, 1:23.200. iTunes automatically selects the Start Time check box for you, so you don't need to select it manually.

 When setting the Start Time value and End Time value, use a colon to separate the minutes and seconds but a period to separate the seconds and thousandths of seconds.

5. Click in the Stop Time box and enter the end time for the ringtone section. Again, iTunes automatically selects the Stop Time check box for you.
6. Click the OK button to close the Item Information dialog box.
7. Right-click or CTRL-click the song, and then click Create AAC Version on the context menu. iTunes creates a new song file containing just the section of the song you specified by using the Start Time value and Stop Time value.

 If the command on the context menu is not Create AAC Version, you need to change the current encoder. Choose Edit | Preferences on Windows or iTunes | Preferences on the Mac to display the iTunes dialog box or the Preferences

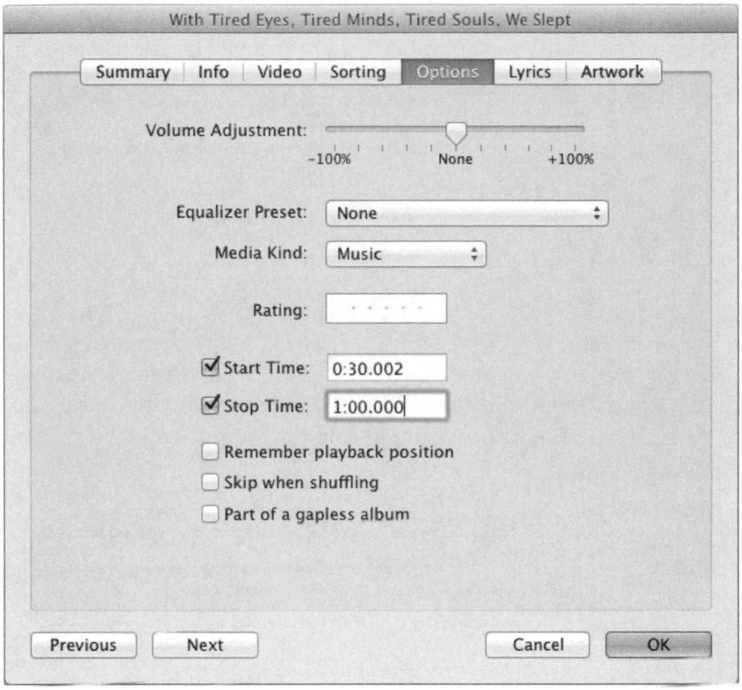

FIGURE 1-12   Use the Item Information dialog box (whose title bar shows "iTunes" on Windows and the song's name on the Mac) to cut a ringtone out of a song.

dialog box. On the General tab, click the Import Settings button. In the Import Settings dialog box, choose AAC Encoder in the Import Using drop-down list and iTunes Plus in the Setting drop-down list. Then click the OK button to close each dialog box in turn.

8. Right-click (or CTRL-click on the Mac) the new, shorter song file, and then click Show In Explorer (on Windows) or Show In Finder (on the Mac) on the context menu. iTunes opens a Windows Explorer window or Finder window showing the song file.

9. Press F2 (on Windows) or RETURN (on the Mac) to display an edit box around the song name.

10. Change the file extension from m4a to m4r, and then press ENTER or RETURN to apply the change. The m4r extension indicates the file type for a ringtone.

11. Leave the Windows Explorer window or Finder window open for the moment and go back to iTunes.

12. With the new song file still selected, choose Edit | Delete. iTunes displays a dialog box confirming you want to remove the file, as shown here.

> **Are you sure you want to remove the selected song from your iTunes library?**
>
> This song will also be removed from any iPod, iPhone, or iPad which synchronizes with your iTunes library.
>
> ☐ Do not ask me again
>
> [Cancel] [Remove]

13. Click the Remove button. iTunes displays a second dialog box asking if you want to move the file to the Recycle Bin (on Windows) or the Trash (on the Mac), as shown here.

> **Do you want to move the selected song to the Trash, or keep it in the iTunes Media folder?**
>
> Only files in the iTunes Media folder will be moved to the Trash.
>
> [Cancel]   [Keep File]   [Move to Trash]

14. Click the Keep File button.
15. In the Windows Explorer window or Finder window, click the ringtone file and drag it to the Library category of the Source list in the iTunes window.
16. You're almost done, but you've set the original song file to play only your ringtone section. Restore it to normality by following these substeps:
    a. Right-click or CTRL-click the original file in the iTunes window, and then choose Get Info to display the Item Information dialog box.
    b. If the Summary tab doesn't appear at the front, click it to bring it there.
    c. Clear the Start Time check box and the Stop Time check box.
    d. Click the OK button to close the Item Information dialog box.
17. Now click the Tones item in the Library category of the Source list to display your ringtones. The file you created appears there, and you can start using it.

Project 6: Spread Your Music Across Your Computers and Devices with iCloud and iTunes Match

Apple's iTunes Match service is a great way of spreading your music automatically among the computers and devices you use. iTunes Match gives you access to online versions of all the songs in your iTunes library. These online versions are stored in iCloud, Apple's new online service.

In order to use iCloud and iTunes Match, your iPhone must be running iOS 5, and preferably the latest version of it. This isn't usually a problem, because Apple has made updating your iPhone easy whether you use iTunes to perform the upgrade or simply upgrade on your iPhone itself.

Understand How iTunes Match Works

To get your music to spread through iCloud, you need to buy an iTunes Match subscription. iTunes Match is Apple's service for giving you access to music online. Here's how iTunes Match works:

- You buy an iTunes Match subscription, which costs $24.99 per year at this writing.
- iTunes then scans all the songs in your music library to see which of them are available in the iTunes Store. The iTunes Store has more than 20 million songs, so chances are that a good proportion of your songs are in it.
- iTunes gives you access to the matching songs in iCloud. These songs are encoded using Advanced Audio Coding (AAC) at the 256 Kbps bitrate, which means they sound good but are compressed small enough for easy streaming.
- iTunes uploads to iCloud all the songs that are in your library but not in iCloud. This takes a while, depending on how many songs are involved and how fast your Internet connection can shift them, but you need to do it only once for each song.

Set Up iTunes Match in iTunes on Your PC or Mac

To set up iTunes Match on your PC or Mac, you use iTunes. Follow these steps:

1. Open iTunes if it's not running, or activate it if it is running.
2. In the Store category in the Source list, click the iTunes Match item to display the iTunes Match screen (see Figure 1-13).
3. Click the Subscribe button. iTunes displays the Sign In To Subscribe To iTunes Match dialog box.
4. Type your password, and then click the Subscribe button. The iTunes Match screen then displays a progress readout (see Figure 1-14) as it goes through the steps of gathering information about your iTunes library, matching your music with songs available in the iTunes Store, and uploading your artwork and unmatched songs.

 The iTunes Match process may take a while. You can stop it if necessary by clicking the Stop button in the lower-right corner of the iTunes Match screen.

5. Use your computer normally while iTunes Match runs.

When iTunes Match finishes running, all the songs in your music library are available to your iPhone, your other iOS devices, and your other computers through iCloud.

FIGURE 1-13   To start setting up iTunes Match, click the iTunes Match item in the Source list, and then click the Subscribe button on the iTunes Match screen.

FIGURE 1-14   iTunes Match goes through the songs in your library, matches as many as possible with songs in the iTunes Store, and uploads your artwork and unmatched songs.

If you stop iTunes Match before it finishes uploading your songs that aren't available in the iTunes Store, iTunes Match restarts automatically each time you launch iTunes. This can come as a surprise, especially as you may find iTunes Match hogging your Internet connection. To turn iTunes Match off until you want to run it again, choose Store | Turn Off iTunes Match.

Turn On iTunes Match on Your iPhone

Now that you've set up your iTunes Match subscription and identified your songs, you can turn on iTunes Match on your iPhone and any other iOS devices.

 Turning on iTunes Match on your iPhone (or other iOS device) replaces the music library on your iPhone. If you prefer to load your iPhone manually with just some of the songs you have in iTunes, don't turn on iTunes Match.

To set up iTunes Match on your iPhone, follow these steps:

1. Press the Home button to display the Home screen.
2. Tap the Settings icon to display the Settings screen.
3. Scroll down to the third box, the one that starts with the General button (shown on the left in Figure 1-15).

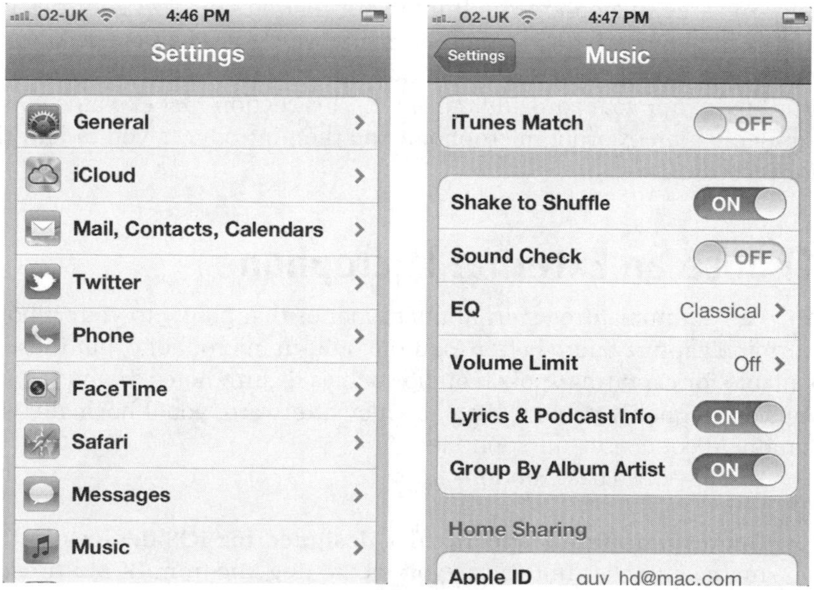

FIGURE 1-15 Tap the Music button on the Settings screen (left) to display the Music screen (right), on which you can turn on iTunes Match by moving the iTunes Match switch to the On position.

4. Tap the Music button to display the Music screen (shown on the right in Figure 1-15).

5. Tap the iTunes Match switch and move it to the On position. Your iPhone displays the Apple ID Password dialog box.

6. Type your password, and then tap the OK button. Your iPhone displays the dialog box shown here, telling you that iTunes Match will replace the music library on your iPhone.

7. Tap the Enable button to turn on iTunes Match.

8. Tap the Settings button to return to the Settings screen.

 As you'd imagine, you set up iTunes Match on an iPod touch in the same way as for your iPhone. The process is the same on the iPad, but you don't need to scroll down.

Project 7: Record High-Quality Audio Using an External Microphone

Your iPhone's built-in microphone is fine for making phone calls, for chatting on video calls via FaceTime, and for recording voice memos using the built-in Voice Memos app. So is the microphone built into your iPhone's headset controls. But if you need to record high-quality audio, you'll normally want to use an external microphone.

If you plan to use an external microphone, you'll normally need to use a third-party app to record audio from it. This section first explains your options for connecting an external microphone and then introduces you to four third-party apps for recording audio.

Choose an External Microphone

You can get miniature external microphones that plug into your iPhone's headphone port and capture audio better than the built-in microphone, but these are mostly suitable for capturing spoken audio such as lecture notes. If you're planning to record music at a quality you'll be able to enjoy afterward, you'll normally want to get a handheld condenser microphone.

You have two basic options here:

- **Get a microphone specifically designed for iOS devices**   At this writing, the main contender in this category is the iRig Mic from IK Multimedia ($59.99; www .ikmultimedia.com and various online stores). The iRig Mic (see Figure 1-16) is a full-size unidirectional condenser microphone with a cable leading to a 3.5-mm jack that connects to your iPhone's headphone socket. The connector also has a headphone socket so that you can listen to the audio.

- **Get a microphone adapter and connect your own microphone** If you want to be able to connect any regular microphone (for example, a high-quality microphone you have already), get a microphone adapter that converts from a 1/4-inch microphone jack or a 1/8-inch microphone jack to your iPhone's 3.5-mm microphone input jack. You can find many such adapters on sites such as Amazon.com and eBay, for prices starting at a handful of dollars. Usually, you'll want to pay enough to get an adapter of a quality at least as high as your microphone so that you don't degrade the signal.

Choose an App for Recording Audio from Your External Microphone

Now that you've chosen your external microphone, you need to get a third-party app that can record audio via that microphone. Here are four of the leading contenders, all of which you can get from the App Store:

FIGURE 1-16 The iRig Mic connects to your iPhone's headphone socket and provides its own headphone socket for monitoring the input. (Photo courtesy of IK Multimedia Production srl.)

- **FiRe (Field Recorder)** If you need to capture live audio, FiRe ($5.99) is a good choice. FiRe can record in either mono or stereo using either your iPhone's built-in microphone or an external microphone that you connect. As you record, FiRe displays a waveform in real time, so you can see what you're getting. The left screen in Figure 1-17 shows FiRe's input screen, on which you can control the gain, choose the quality, decide whether to play audio through, turn audio processing on or off, and choose which preset to use. You can choose among different presets, such as Male Voice Enhancer, Female Voice Enhancer, Live Concert Outdoors, and Noise Gate. The right screen in Figure 1-17 shows FiRe in action recording audio.
- **iRig Recorder** If you went with the iRig Mic as your microphone, iRig Recorder may seem the obvious choice as your recording app. Your best approach is to start with the free version, iRig Recorder FREE, and then either buy the full version of iRig Recorder for $4.99 or buy only the add-ons you want. For example, you may want to buy the editing add-on but not the processing add-on.

FIGURE 1-17   After choosing options such as Gain, Quality, and Playthrough on the Input screen (left), you can set FiRe recording (right) and see the waveform of the audio you're capturing.

- **ISW Recorder and Editor**   ISW Recorder and Editor is free, so it's well worth trying to see if it meets your needs. You can cut recordings down to only the parts you need, rearrange audio snippets into your preferred order, and share them via e-mail, Twitter, or Facebook. The left screen in Figure 1-18 shows ISW Recorder and Editor.
- **iProRecorder**   iProRecorder ($4.99) is a business-oriented recorder intended mainly for dictation and transcription, although you can of course use it to record any other audio as well. iProRecorder (shown on the right in Figure 1-18) features adjustable playback speed and a jog/shuttle wheel, both of which are helpful when you're transcribing a recording.

Project 8: Play Your Guitar Through Your iPhone

If you play electric guitar, you can connect it to your iPhone and play it through your iPhone. This is great because you can not only play your guitar through headphones so that you blast your own ears rather than the neighbors, but also use your iPhone as a bunch of effects pedals to get the sound you want. Your iPhone is easier to carry than a bagful of effects pedals, and the apps for producing the effects cost much less than the physical pedals.

FIGURE 1-18   ISW Recorder and Editor (left) is a free recorder that includes basic editing capabilities. iProRecorder (right) is a business-oriented recorder that features adjustable playback speed and a jog/shuttle wheel for making transcription easier.

 You can use this technique for any instrument that has a pickup—electric bass, electric violin, or whatever.

In this section, we'll first get your guitar connected to your iPhone with a cable. We'll then look at special-effects apps you can use to enhance the sound.

Connect Your Guitar to Your iPhone

To connect your guitar to your iPhone, you'll need either a cable that goes from your guitar's 1/4-inch output to your iPhone's headphone port or an adapter that lets you make this connection. Here are two of the leading possibilities:

- **GuitarConnect Cable**   The GuitarConnect Cable from Griffin Technology ($29.99; www.griffintechnology.com) is a guitar cable with a built-in splitter. You plug the GuitarConnect's 1/4-inch jack into your guitar, plug the 1/8-inch jack on the other end into your iPhone's headphone socket, and optionally plug your headphones into the headphone port on the GuitarConnect.

- **AmpliTube iRig** The AmpliTube iRig from IK Multimedia ($39.99; www
.ikmultimedia.com or sites such as Amazon.com) is a guitar connector and splitter.
You plug your regular guitar lead into one end of the iRig (see Figure 1-19), plug
the cable at the other end into your iPhone's headphone socket, and optionally
plug your headphones into the other port on the iRig.

 You can also connect the headphone port on the GuitarConnect Cable or the
AmpliTube iRig to an amplifier or stereo.

Now that you've connected your guitar, what you play goes into your iPhone,
where you can record it or run it through an effects app, as discussed next.

Apply Special Effects to Your Guitar

Now that your guitar input is going into your iPhone, you can apply effects to it by
using an app such as one of these:

- **AmpliTube** AmpliTube is a family of effects apps from IK Multimedia designed
to work with the iRig. There are enough versions to be confusing. You'll probably
want to start with AmpliTube FREE or AmpliTube Fender FREE before moving
on to AmpliTube ($19.99), AmpliTube Fender ($14.99), or AmpliTube LE ($2.99).
Figure 1-20 shows AmpliTube.

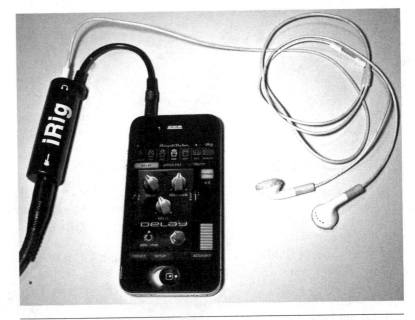

FIGURE 1-19 The AmpliTube iRig gives you an easy way to connect
your guitar to your iPhone's headphone socket. You can also plug your
headphones into the iRig to listen to what you're playing.

FIGURE 1-20   AmpliTube is a family of effects apps that works with the iRig guitar connector.

- **iShred LIVE**   iShred LIVE is a stompbox effects app. iShred LIVE is free, but you have to pay for effects—Adrenaline, Kömpressör, Trembler, Screamer, Octavinator, and so on. Most effects cost $0.99 each, but some cost more. You can also buy a Power Pack that contains all the effects. Figure 1-21 shows iShred LIVE.

Having your iPhone do the work of effects pedals is great, but it means that you need to tap your iPhone's screen to change effects. If you want to be able to change effects without interrupting your playing, consider getting the StompBox controller from Griffin Technology ($99.99; www.griffintechnology.com or various online retailers). The StompBox (shown in Figure 1-22 connected to an iPad, which it works with as well) is a physical pedal that you can connect to your iPhone and use to control the effects in iShred LIVE.

 To keep your effects at your fingertips, get a case or holster that lets you clip your iPhone to your belt. Alternatively, adapt a case to mount your iPhone on your guitar, where you can reach it easily.

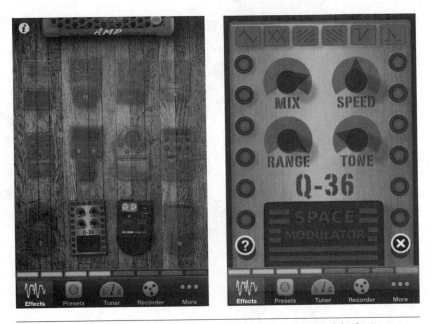

FIGURE 1-21 iShred LIVE is an effects app that works with the GuitarConnect Cable.

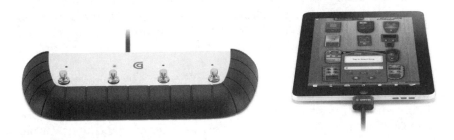

FIGURE 1-22 Add the Griffin StompBox to your iPhone or iPad guitar setup if you want to control your effects with your feet as you play. (Photo courtesy of Griffin Technology, Inc.)

Project 9: Record Your Band on Your iPhone

If you play live in a band, you'll probably want to record it. Your iPhone is a great tool for the task once you equip it with suitable hardware and software.

 You can record audio using the Voice Memos app and your iPhone's built-in microphone, but you will normally get better results by using an external microphone and a third-party recording app that lets you choose the settings you need. See Project 7, "Record High-Quality Audio Using an External Microphone," for suggestions on which microphone to choose.

In this section, you'll first choose how to input your audio into your iPhone. You'll then choose a recording app that can capture the audio.

Choose Your Input

If you want to record a live performance, you can simply use a microphone as discussed in Project 7, "Record High-Quality Audio Using an External Microphone," and an app such as iRig Recorder or FiRe. But you can also use your iPhone as a multi-track recorder by installing the right app. You can then lay down a single track at a time, just as you would with a physical multi-track recorder, and mix the tracks together to produce the result you want.

To capture input from a voice or from an acoustic instrument without a pickup, use a microphone as mentioned above, but record it as a track on a multi-track recorder.

To capture input directly from a "real" instrument such as a guitar or bass, connect it using a cable such as the Griffin GuitarConnect Cable or the AmpliTube iRig. See the previous project for details on these connectors.

To capture MIDI input from a keyboard, drum machine, or other device that has a MIDI output, get a MIDI interface such as the iRig MIDI ($69.99; www.ikmultimedia.com) or the MIDI Mobilizer ($99.99 list price, but widely available for much less; http://line6.com/midimobilizer/).

Choose a Recording App

What you need next is a suitable app for recording the audio that the microphone or input picks up. Here are four apps to consider:

- **Multi Track Song Recorder**   Multi Track Song Recorder (shown on the left in Figure 1-23) is a free recorder that can record up to four tracks. You can view the tracks as either waveforms or as volume meters, and you can set the volume of each track to a suitable level to create the overall mix. You can import songs from your iPhone's music library into tracks, which can be a great way of getting a song started quickly. The right screen in Figure 1-23 shows a command dialog box in Multi Track Song Recorder.

FIGURE 1-23   Multi Track Song Recorder is free—supported by ads—and can record up to four tracks.

- **FiRe Studio**   FiRe Studio ($4.99) can record and mix up to eight tracks, giving you plenty of flexibility. You can scroll quickly through the waveforms, place the playback head where you want to start playback, and lock finished tracks to prevent changes. Figure 1-24 shows FiRe Studio.
- **StudioApp**   StudioApp ($4.99) is a recorder that enables you to add up to four tracks to instrumental tracks. StudioApp (shown in Figure 1-25) targets hip-hop artists, rappers, and singers, but it works for any audio you input.

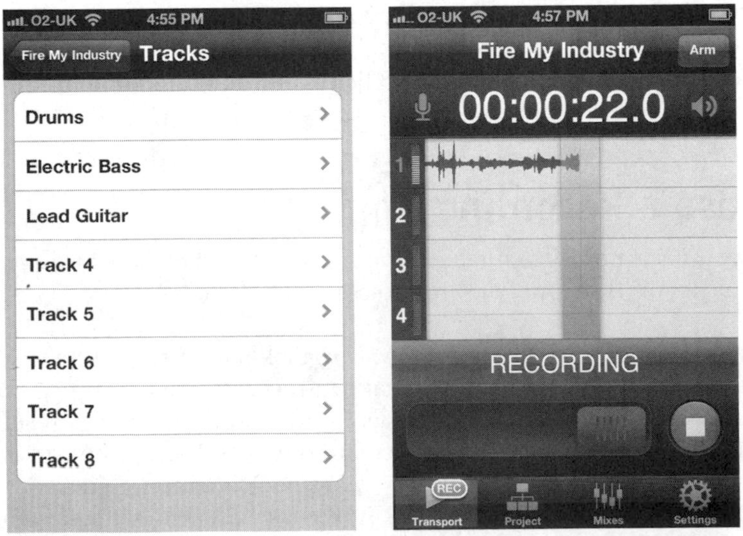

FIGURE 1-24   FiRe Studio costs $4.99 and can record and mix up to eight tracks.

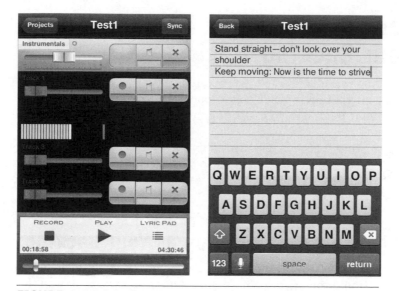

FIGURE 1-25   StudioApp lets you add up to four tracks to instrumental tracks. You can use the built-in Lyric Pad to capture your ideas.

- **VocaLive**   If you want to record and process vocals, try VocaLive (see Figure 1-26). Start with the free version, VocaLive Free, and graduate to the paid version (simply called VocaLive and costing $19.99) if it suits you. The full version includes a real-time vocal processor and a dozen vocal effects—including Pitch Fix, Choir, De-Esser, and Chorus—that enable you to make your vocals sound substantially different (better, unless you prefer it otherwise).

FIGURE 1-26   VocaLive is an app for recording and processing vocals.

Project 10: Use Your iPhone as Your Backing Track

Playing music along with your band is great, but chances are you'll sometimes need to play on your own. When this happens, you don't need to play alone, because you can use your iPhone as your backing track.

In this section, I'm assuming you want to play music you've created as your backing track. If you want to play along to someone else's music, you can pick up karaoke mixes or guitar-free mixes of many songs from various sites on the Internet.

Choose a Suitable App as Your Backing Track

Which app will suit you best to provide your backing track depends on what you need to do—but here are three examples of apps that you may want to look at:

- **GigBaby**   GigBaby ($0.99) is a four-track recorder with a built-in rhythm section. You can set the rhythm you want, record other backing tracks to play along with it, and then use GigBaby either as your accompaniment or to record your lead performance. Figure 1-27 shows GigBaby at work.

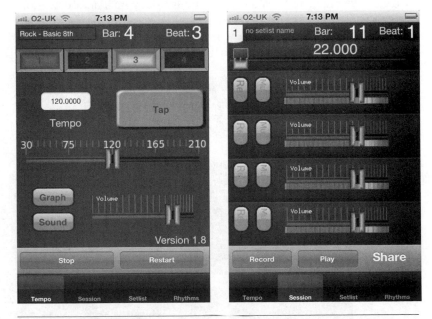

FIGURE 1-27   GigBaby is a four-track recorder with built-in rhythms you can use as a base for your songs.

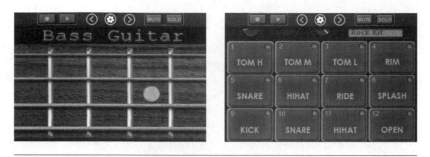

FIGURE 1-28 Band provides a set of virtual instruments that you can either play live or record as your accompaniment.

- **Band** Band ($3.99) is an app for playing virtual instruments onscreen—bass, grand piano, and two drum kits (see Figure 1-28). You can play the instruments in real time at a pinch, but what you'll normally want to do is record your instrumental parts so that Band can play them back while you play along on a real instrument.
- **BeatMaker** BeatMaker ($9.99) and BeatMaker 2 ($19.99) are high-powered sequencing apps. You can load an existing kit or develop a custom kit, play it live or record it, and arrange patterns into tracks that sound the way you want them to. Figure 1-29 shows the pads in a BeatMaker kit loaded and ready to play or record.

FIGURE 1-29 BeatMaker is a sequencing app that lets you play existing kits or develop custom kits containing the sounds you want.

Connect Your iPhone to Your Amp or Sound Board

To get your backing track to play along with you, you'll need to connect your iPhone to your amp or to your sound board.

You can connect your iPhone via its headphone socket using a cable with a 3.5-mm plug at the iPhone's end and whatever is needed at the amplifier's end—for example, a 1/4-inch plug or two RCA plugs. But if you have the choice, use a cable that has a Dock Connector at the iPhone's end. Using the iPhone's Dock Connector port gives you a line-level output, which has a constant volume and is much easier to work with than the output of the headphone port, whose volume depends on the volume setting.

 If you have an iPhone dock that has a line-out port, you can use that instead of getting a cable with a Dock Connector.

2 Photo and Video Geekery

Equipped with its high-resolution camera at the back and the user-facing camera at the front, your iPhone is ready to capture video either fore or aft. It's also pretty good at playing back video, either on its screen for your enjoyment or on an external monitor or a TV so that you can share it with others without necessarily getting intimate.

We'll start this chapter by looking at how to put your videos and DVDs on your iPhone so you can watch them wherever you want. We'll then move on to the procedure for watching video from your iPhone on your TV. We'll then look at how to share your photos among your iPhone and your other devices by using Apple's Photo Stream feature.

After that, I'll show you how to take macro and panorama photos with your iPhone's camera, how to take high-quality self-portraits, and how to take time-lapse movies and shoot video at different frame rates.

Toward the end of the chapter, I'll show you how to build your own Steadicam rig to hold and stabilize your iPhone so that you can shoot good-quality video while you're moving. And last, we'll go through how to view your webcam on your iPhone, either as a party trick (if you have that kind of party) or to keep tabs on home when you're away.

Project 11: Put Your Videos and DVDs on Your iPhone

Apple's iTunes Store provides a wide selection of video content, including TV series and full-length movies, and you can buy or download video in iPhone-compatible formats from various other sites online.

But if you enjoy watching video on your iPhone, you'll almost certainly want to put your own video content on it. You may also want to rip files from your own DVDs so that you can watch them on your iPhone. This project shows you how to do so.

Create iPhone-Friendly Video Files from Your Digital Video Camera

If you make your own movies with a digital video camera, you can easily put them on the iPhone. To do so, you use an application such as Windows Movie Maker (Windows) or iMovie (Mac) to capture the video from your digital video camera and turn it into a home movie.

 Video formats are confusing at best—but the iPhone and iTunes make the process of getting suitable video files as easy as possible. The iPhone can play videos in the MP4 format up to 2.5 Mbps (megabits per second) or the H.264 format up to 720p. Programs designed to create video files suitable for the iPhone typically give you a choice between the MP4 format and the H.264 format. As a point of reference, VHS video quality is around 2 Mbps, while DVD is about 8 Mbps.

Create iPhone-Friendly Video Files Using Windows Live Movie Maker or Windows Movie Maker

Unlike the last few versions of Windows, Windows 7 doesn't include Windows Movie Maker, the Windows program for editing videos. But you can download the nearest equivalent, Windows Live Movie Maker, from the Windows Live website (http://explore.live.com/windows-live-movie-maker?os = other).

 When you install Windows Live Movie Maker, the Windows Live Essentials installer encourages you to install all the Windows Live Essentials programs—Messenger, Photo Gallery, Mail, Writer, Family Safety, and several others. If you don't want the full set, click the Choose The Programs You Want To Install button on the What Do You Want To Install? screen, and then select only the programs you actually want.

 DOUBLE GEEKERY

Learn What You Can and Can't Legally Do with Other People's Video Content

Before you start putting your videos and DVDs on the iPhone, it's a good idea to know the bare essentials about copyright and decryption:

- If you created the video (for example, it's a home video or DVD), you hold the copyright to it, and you can do what you want with it—put it on the iPhone, release it worldwide, or whatever. The only exceptions are if what you recorded is subject to someone else's copyright or if you're infringing on your subjects' rights (for example, to privacy).
- If someone has supplied you with a legally created video file that you can put on your iPhone, you're fine doing so. For example, if you download a video from the iTunes Store, you don't need to worry about legalities.
- If you own a copy of a commercial DVD, you need permission to rip (extract) it from the DVD and convert it to a format the iPhone can play. Even decrypting the DVD in an unauthorized way (such as creating a file rather than simply playing the DVD) is technically illegal.

Windows Live Movie Maker can't export video files in an iPhone-friendly format, so what you need to do is export the video file in the WMV format, and then convert it using another application, such as Full Video Converter Free (discussed later in this chapter).

Similarly, the versions of Windows Movie Maker included with Windows Vista and Windows XP can't export video files in an iPhone-friendly format, so what you need to do is export the video file in a standard format (such as AVI) that you can then convert using another application.

Create a WMV File from Windows Live Movie Maker   To create a WMV file from Windows Live Movie Maker, open the project and follow these steps:

1. Choose File | Save Movie to display the Save Movie panel. The tab I'm calling "File" here is the unnamed tab at the left end of the Ribbon.
2. In the Common Settings section, click For Computer. The Save Movie dialog box opens.
3. Type the name for the movie, choose the folder in which to store it, and then click the Save button.

Now that you've created a WMV file, use a converter program such as Full Video Converter Free (discussed later in this chapter) to convert it to a format that the iPhone can play.

Create an AVI File from Windows Movie Maker on Windows Vista   To save a movie as an AVI file from Windows Movie Maker on Windows Vista, follow these steps:

1. With your movie open in Windows Movie Maker, choose File | Publish Movie (or press CTRL-P) to launch the Publish Movie Wizard. The Wizard displays the Where Do You Want To Publish Your Movie? screen.
2. Select the This Computer item in the list box, and then click the Next button. The Wizard displays the Name The Movie You Are Publishing screen.
3. Type the name for the movie, choose the folder in which to store it, and then click the Next button. The Wizard displays the Choose The Settings For Your Movie screen (see Figure 2-1).
4. Select the More Settings option button, and then select the DV-AVI item in the drop-down list.

 The DV-AVI item appears as DV-AVI (NTSC) or DV-AVI (PAL), depending on whether you've chosen the NTSC option button or the PAL option button on the Advanced tab of the Options dialog box. NTSC is the video format used in most of North America; PAL's stronghold is Europe.

5. Click the Publish button to export the movie in this format. When Windows Movie Maker finishes exporting the file, it displays the Your Movie Has Been Published screen.

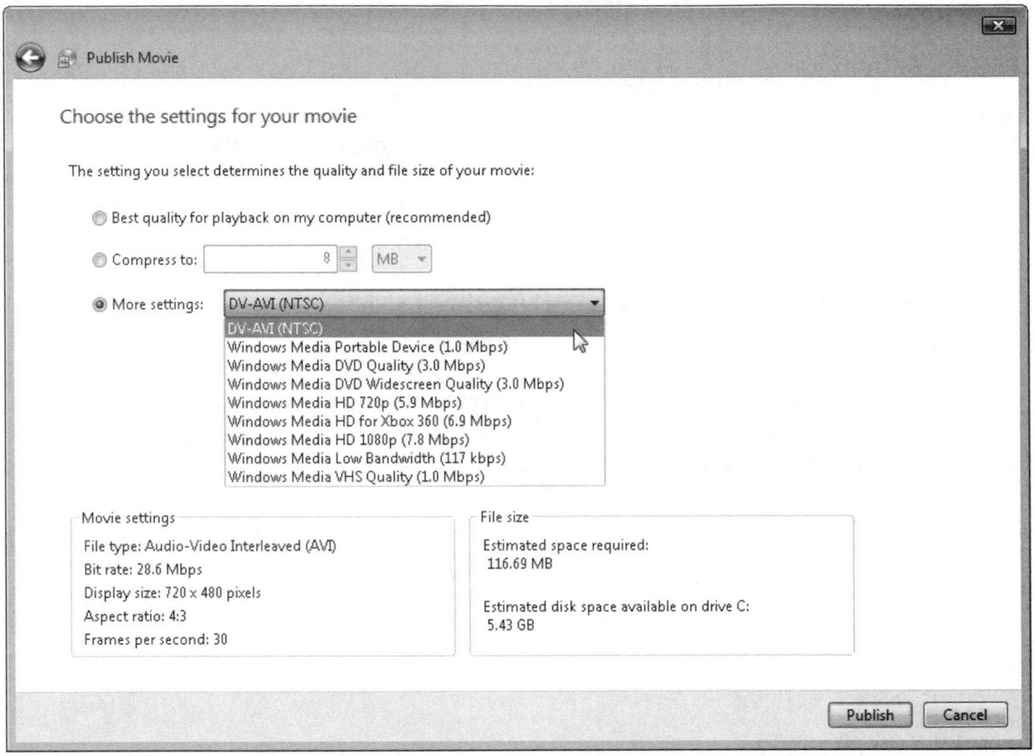

FIGURE 2-1   On the Choose The Settings For Your Movie screen, select the More Settings option button, and then pick the DV-AVI item in the drop-down list.

6. Clear the Play Movie When I Click Finish check box if you don't want to watch the movie immediately in Windows Media Player. Often, it's a good idea to check that the movie has come out okay.
7. Click the Finish button.

Now that you've created an AVI file, use a converter program such as Full Video Converter Free (discussed later in this chapter) to convert it to a format that works on the iPhone.

Create an AVI File from Windows Movie Maker on Windows XP   To save a movie as an AVI file from Windows Movie Maker on Windows XP, follow these steps:

1. Choose File | Save Movie File to launch the Save Movie Wizard. The Wizard displays its Movie Location screen.
2. Select the My Computer item, and then click the Next button. The Wizard displays the Saved Movie File screen.

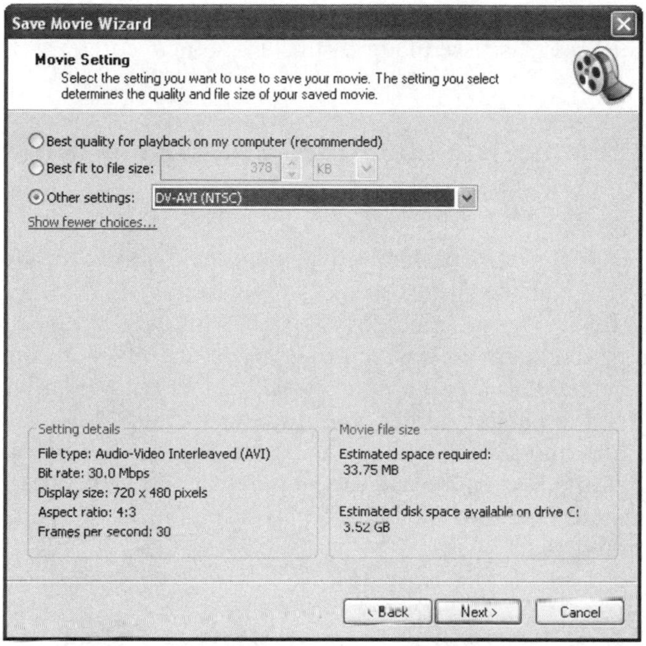

Save Movie Wizard

Movie Setting
Select the setting you want to use to save your movie. The setting you select
determines the quality and file size of your saved movie.

○ Best quality for playback on my computer (recommended)
○ Best fit to file size: 370   KB
◉ Other settings: DV-AVI (NTSC)

Show fewer choices...

Setting details

File type: Audio-Video Interleaved (AVI)
Bit rate: 30.0 Mbps
Display size: 720 x 480 pixels
Aspect ratio: 4:3
Frames per second: 30

Movie file size

Estimated space required:
33.75 MB

Estimated disk space available on drive C:
3.52 GB

〈 Back   Next 〉   Cancel

FIGURE 2-2   Click the Show More Choices link to make
the Other Settings option button available, then select the
Other Settings option button and pick the DV-AVI item from
the drop-down list.

3. Enter the name and choose the folder for the movie, and then click the Next
button. The Wizard displays the Movie Setting screen (shown in Figure 2-2 with
options selected).
4. Click the Show More Choices link to display the Best Fit To File Size option
button and the Other Settings option button.
5. Select the Other Settings option button, and then select the DV-AVI item in the
drop-down list.

 The DV-AVI item appears as DV-AVI (NTSC) or DV-AVI (PAL), depending on whether
you've chosen the NTSC option button or the PAL option button on the Advanced
tab of the Options dialog box. NTSC is the video format used in most of North
America; PAL's stronghold is Europe.

6. Click the Next button to save the movie in this format. The Wizard displays the
Completing The Save Movie Wizard screen.
7. Clear the Play Movie When I Click Finish check box if you don't want to test
the movie immediately in Windows Media Player. Usually, it's a good idea to
make sure the movie has come out right.
8. Click the Finish button.

Now that you've created an AVI file, use a converter program such as Full Video Converter Free (discussed later in this chapter) to convert it to a format that works on the iPhone.

Create iPhone-Friendly Video Files Using iMovie

To use iMovie to create video files that will play on the iPhone, follow these steps:

1. With the movie open in iMovie, choose Share | iTunes to display the Publish Your Project To iTunes sheet (see Figure 2-3).
2. In the Sizes area, select the check box for each size you want to create. The dots show the devices for which that size is suitable. For example, if you want to play the video files on an iPhone classic, select the Medium check box.
3. Click the Publish button, and then wait while iMovie creates the compressed file or files and adds it or them to iTunes. iMovie then automatically displays iTunes.
4. Click the Movies item in the Source list, and you'll see the movies you just created. Double-click a file to play it, or simply drag it to the iPhone to load it immediately.

Create iPhone-Friendly Video Files from Your Existing Video Files

If you have existing video files (for example, files in the AVI format or QuickTime movies), you can convert them to iPhone format in a couple of ways. The easiest way is by using the capabilities built into iTunes—but unfortunately, these work only for some video files. The harder way is by using QuickTime Pro, which can convert files from most known formats but which costs $30.

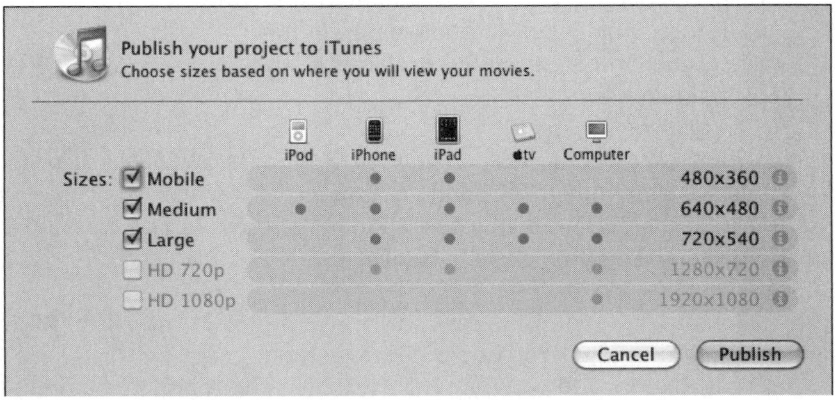

FIGURE 2-3   On the Publish Your Project To iTunes sheet in iMovie, choose which sizes of file you want to create—for example, Medium for the iPhone.

On Windows, you can also use third-party converter programs, such as Full Video Converter Free, discussed later in this chapter.

Create iPhone-Friendly Video Files Using iTunes

To create a video file for the iPhone using iTunes, follow these steps:

1. Add the video file to your iTunes library in either of these ways:
   - Open iTunes if it's not running. Open a Windows Explorer window (Windows) or a Finder window (Mac) to the folder that contains the video file. Arrange the windows so that you can see both the file and iTunes. Drag the file to the Library item in iTunes.
   - In iTunes, choose File | Add To Library, use the Add To Library dialog box to select the file, and then click the Open button (Windows) or the Choose button (Mac).
2. Select the movie in the iTunes window, and then choose Advanced | Create iPhone Or iPod Version.

If the Create iPhone Or iPod Version command isn't available for the file, or if iTunes gives you an error message, you'll know that iTunes can't convert the file.

Create iPhone-Friendly Video Files Using QuickTime

QuickTime, Apple's multimedia software for Mac OS X and Windows, comes in two versions: QuickTime Player (the free version) and QuickTime Pro, which costs $29.99.

Create iPhone-Friendly Video Files Using QuickTime Player on the Mac   On Mac OS X, QuickTime Player is included in a standard installation of the operating system; and if you've somehow managed to uninstall it, it'll automatically install itself again if you install iTunes. The Mac version of QuickTime Player includes file conversions, which you can access by using the Share menu. For example, follow these steps:

1. Open QuickTime Player from Launchpad, the Dock, or the Applications folder.
2. Choose File | Open File, select the file in the Open dialog box, and then click the Open button.
3. Choose Share | iTunes to display the Save Your Movie To iTunes dialog box (see Figure 2-4).
4. Select the iPod & iPhone option button.
5. Click the Share button. QuickTime converts the file.

Create iPhone-Friendly Video Files Using QuickTime Pro on Windows   On Windows, you install QuickTime Player when you install iTunes, because QuickTime provides much of the multimedia functionality for iTunes. The "Player" name isn't entirely accurate, because QuickTime provides encoding services as well as decoding services

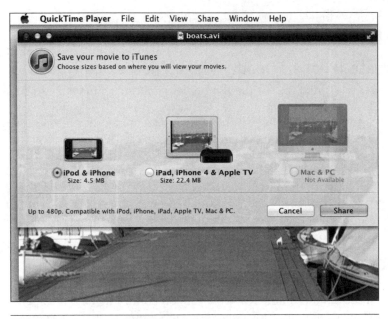

FIGURE 2-4   On the Mac, you can use QuickTime Player to convert
video files to formats suitable for the iPhone.

to iTunes—but QuickTime Player on the PC doesn't allow you to create most formats of
video files until you buy QuickTime Pro.

 QuickTime Pro for Windows gets rave reviews from some users but wretched reviews
from others. If you are thinking of buying QuickTime Pro for Windows, read the
latest reviews for it at the Apple Store (http://store.apple.com) first.

QuickTime Player for Windows is a crippled version of QuickTime Pro, so when
you buy QuickTime Pro from the Apple Store, all you get is a registration code to unlock
the hidden functionality. To apply the registration code, choose Edit | Preferences |
Register In Windows to display the Register tab of the QuickTime Settings dialog box.
On the Mac, choose QuickTime Player | Registration to display the Register tab of the
QuickTime dialog box.

 When you register QuickTime Pro, you must enter your registration name in the
Registered To text box in exactly the same format as Apple has decided to use it. For
example, if you've used the name John P. Smith to register QuickTime Pro, and Apple
has decided to address the registration to *Mr. John P. Smith*, you must use **Mr. John
P. Smith** as the registration name. If you try to use **John P. Smith**, registration fails,
even if this is exactly the way you gave your name when registering.

To create an iPhone-friendly video file from QuickTime Pro, follow these steps:

1. Open the file in QuickTime Pro, and then choose File | Export to display the Save Exported File As dialog box.
2. Specify the filename and folder as usual, and then choose Movie To iPhone in the Export drop-down list. Leave the Default Settings item selected in the Use drop-down list.
3. Click the Save button to start exporting the video file.

Create iPhone-Friendly Video Files Using Full Video Converter Free

If you have video files that you can't convert with iTunes on Windows, use a file conversion program such as Full Video Converter Free (see Figure 2-5). You can download this program from Top 10 Download (www.top10download.com) and other sites. When you install the program, make sure you decline any extra options such as adding a toolbar, changing your default search engine, or changing your home page.

 You can find various other free programs online for converting video files. If you're looking for such programs, check carefully that what you're about to download is actually free rather than a crippled version that requires you to pay before you can convert files.

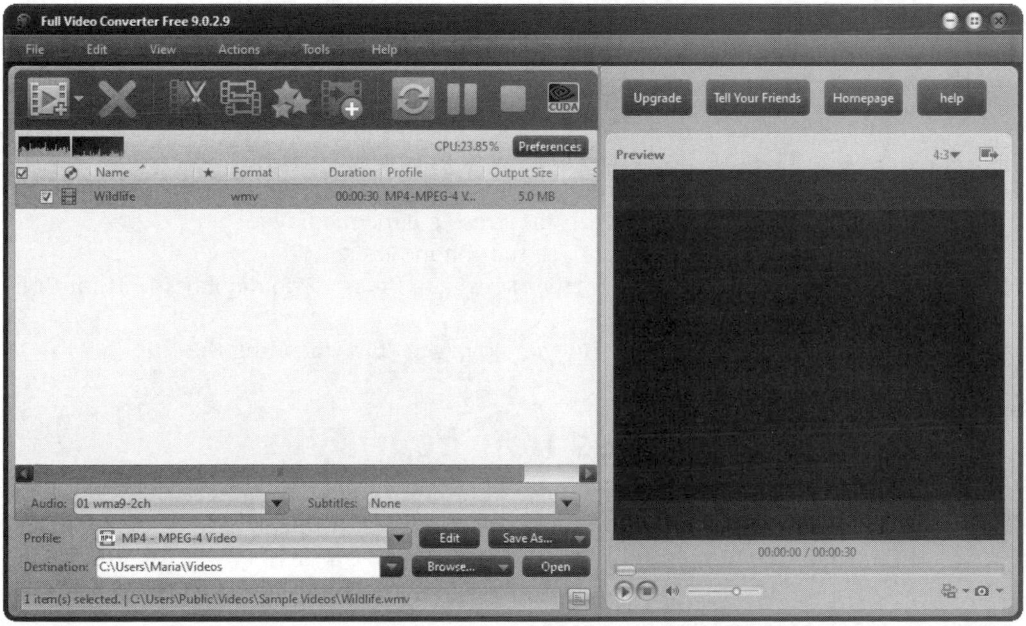

FIGURE 2-5 Full Video Converter Free lets you convert various types of videos to iPhone-friendly formats.

Another way to convert video files from one format to another—on either Windows or the Mac—is to use an online file conversion tool such as Zamzar (www.zamzar .com). For low volumes of files, the conversion is free (though it may take a while), but you must provide a valid e-mail address. For higher volumes of files or higher priority, you can sign up for a paid account.

Create iPhone-Friendly Video Files Using HandBrake on the Mac

If you have video files that you can't convert with iTunes on the Mac, try using the free conversion program HandBrake (http://handbrake.fr). Download HandBrake, install it to your Applications folder, run it from there, and then follow these steps:

1. Click the Source button on the toolbar to display an Open dialog box.

 HandBrake can also rip DVDs, provided you have a third-party decryption utility installed. See the end of this section for details.

2. Click the file you want, and then click the Open button. HandBrake shows the details of the file.
3. In the Title drop-down list, choose which title—which of the recorded items in the file—you want. Most files have only one title, so the choice is easy; DVDs have various titles.
4. If the file is broken up into chapters (sections), choose which ones you want. Pick the first in the Chapters drop-down list and the last in the Through drop-down list—for example, Chapters 1 through 4.
5. In the Destination area, change the name and path for the converted file if necessary.
6. If the Presets pane isn't displayed on the right side of the window, click the Toggle Presets button on the toolbar to display it. Figure 2-6 shows the HandBrake window with the Presets pane displayed.
7. In the Presets pane, choose the iPhone preset.
8. If necessary, change further settings. (Press ⌘-? to display the HandBrake User Guide for instructions.)
9. Click the Start button on the toolbar to start encoding the file.

Create Video Files from Your DVDs

If you have DVDs, you'll probably want to put them on the iPhone so that you can watch them without a DVD player. This section gives you an overview of how to create suitable files, first on Windows, and then on the Mac.

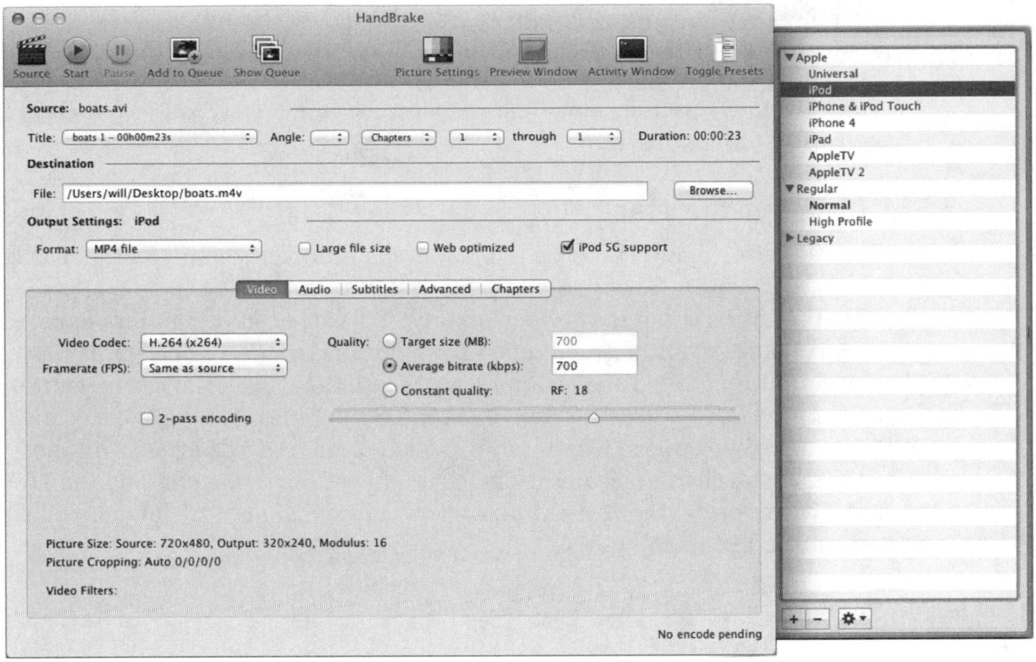

FIGURE 2-6   The Presets pane on the right side of the HandBrake window lets you instantly choose video settings for the iPhone.

Because ripping commercial DVDs without specific permission is a violation of copyright law, there are no DVD ripping programs from major companies. You can find commercial programs, shareware programs, and freeware programs on the Internet—but keep your wits firmly about you, as some programs are a threat to your computer through being poorly programmed, while others include unwanted components such as adware or spyware. Always read reviews of any DVD ripper you're considering before you download and install it—and certainly before you pay for it. As usual on the Internet, if something seems too good to be true, it most likely *is* too good to be true.

Before you start ripping, make sure that your discs don't contain computer-friendly versions of their contents. At this writing, some Blu-Ray discs include such versions, which are licensed for you to load on your computer and your lifestyle devices (such as the iPhone).

Rip DVDs on Windows

Here are two solutions for decrypting and ripping DVDs on Windows:

- **DVD43 and DVD Shrink**   DVD43 is a free DVD-decryption utility that you can download from the links on the DVD43 – Download Sites page (www.dvd43.com). DVD43 opens the DVD up for ripping but doesn't rip the content from the DVD. To rip, use a program such as DVD Shrink ($28.95; www.official-dvdshrink.org).

- **AnyDVD and CloneDVD Mobile** AnyDVD from SlySoft (around $55 per year; www.slysoft.com) is a decryption utility that works with CloneDVD Mobile (around $45 per year; also from SlySoft). By using these two programs together, you can rip DVDs to formats that work on the iPhone. SlySoft offers 21-day trial versions of these programs.

Rip DVDs on the Mac

The best tool for ripping DVDs on the Mac is HandBrake, which you met earlier in this chapter. To rip DVDs with HandBrake, you must install VLC, a DVD- and video-playing application (free; www.videolan.org). This is because HandBrake uses VLC's DVD-decryption capabilities; without VLC, HandBrake cannot decrypt DVDs.

Once you've installed VLC, simply run HandBrake, click the Source button, click the DVD in the Source list, and then click the Open button. HandBrake scans the DVD. You can then choose which "title" (which of the recorded tracks on the DVD) to rip, and which chapters from it. The chapters are the bookmarks on the DVD—for example, if you press the Next button on your remote, your DVD player skips to the start of the next chapter.

DOUBLE GEEKERY

Prevent the Mac OS X DVD Player from Running Automatically When You Insert a DVD

When you insert a movie DVD, Mac OS X automatically launches DVD Player, switches it to full screen, and starts the movie playing. This behavior is great for when you want to watch a movie, but not so great when you want to rip it.

To prevent DVD Player from running automatically when you insert a DVD, follow these steps:

1. Choose Apple | System Preferences to open System Preferences.
2. In the Hardware section, click the CDs & DVDs item.
3. In the When You Insert A Video DVD drop-down list, you can choose Ignore if you want to be able to choose freely which application to use each time. If you always want to use the same application, choose Open Other Application, use the resulting Open dialog box to select the application, and then click the Choose button.
4. Choose System Preferences | Quit System Preferences or press ⌘-Q to close System Preferences.

Project 12: Watch Video from Your iPhone on a TV

Once you've loaded your video files onto your iPhone, you're ready to watch them anywhere. Watching on your iPhone's screen works fine when you yourself are the whole audience, but when you need to share your videos with other people, you'll likely want a bigger screen. Often, the easiest solution is to play video from your iPhone to a TV.

Connect Your iPhone to the TV

To play videos from your iPhone on a TV, you need a suitable cable. Look first at the Apple Composite AV Cable and the Apple Component AV Cable on the Apple Store (http://store.apple.com) and establish which one your TV needs. Then decide between buying the Apple version of the cable or a third-party equivalent.

When you have the cable, connect it to your iPhone's Dock Connector port and to the appropriate inputs on your TV. You know what the iPhone's end of the connection looks like.

Play Back a Video or Movie on the TV

After connecting your iPhone to the TV, you can play back a video or movie on the TV by simply starting playback on the iPhone as usual.

When you start playback, your iPhone tells you that the output is going to the TV, as shown here.

 If the TV isn't showing the video, you'll need to fiddle with the AV buttons to make sure it's using the right input.

When you finish watching the video, disconnect the cable from the TV.

Project 13: Share Your Photos with All Your Devices Using Photo Stream

Being able to take high-quality photos anywhere with your iPhone's built-in camera is great. But what's even better is being able to use the Photo Stream feature to make those photos appear on your computer and your other iOS devices (for example, your iPad) automatically.

In this section, I'll show you how to set up and use Photo Stream.

Understand What Photo Stream Is and What It Does

Photo Stream is part of Apple's iCloud service, so to use it you must have an iCloud account. Given that you've got an iPhone, you've probably set up an iCloud account already; if not, you can set one up inside a couple of minutes.

Once you've set it up, Photo Stream automatically syncs up to 1,000 of your latest photos among your iOS devices and your computers. Photo Stream stores each new photo on iCloud for 30 days, so if you connect each iOS device to a wireless network several times a week, you'll soon have each new photo on each device.

On your iPhone (or your iPod touch, or your iPad), Photo Stream includes the photos in your Camera Roll folder. The Camera Roll contains not only the photos you take using the Camera app but also photos you save from e-mail messages, multimedia messages, or web pages.

Photo Stream works with any device running iOS 5—the iPhone (3GS, 4, or 4S), any iPad, or the iPod touch (third-generation or later). It works with iPhoto or Aperture on the Mac and with the Pictures Library on Windows 7 or Windows Vista.

Set Up Photo Stream on Your iPhone or iPod touch

To set up Photo Stream on your iPhone or iPod touch, follow these steps:

1. Press the Home button to display the Home screen.
2. Tap the Settings icon to display the Settings screen.
3. Scroll down to the third box, the one that starts with the General button, and then scroll down a bit further, so that you see the apps shown on the left in Figure 2-7.

FIGURE 2-7 Tap the Photos button on the Settings screen to display the Photos screen (right), and then move the Photo Stream switch to the On position.

4. Tap the Photos button to display the Photos screen (shown on the right in Figure 2-7).
5. Tap the Photo Stream switch and move it to the On position.

When you set the Photo Stream switch to the On position, your iPhone or iPod touch prompts you to sign into iCloud if you're not currently signed in.

6. Tap the Settings button to go back to the Settings screen.

Set Up Photo Stream on Your iPad

To set up Photo Stream on your iPad, follow these steps:

1. Press the Home button to display the Home screen.
2. Tap the Settings icon to display the Settings screen.
3. Tap the Photos button in the left column to display the Photos screen.
4. Tap the Photo Stream switch and move it to the On position.

When you set the Photo Stream switch to the On position, your iPad prompts you to sign into iCloud if you're not currently signed in.

Set Up Photo Stream on Your PC

If you have a PC running Windows 7 or Windows Vista, you can set up Photo Stream to sync your photos automatically. To do so, you install the iCloud Control Panel, then log in to your iCloud account and turn on Photo Stream. You can also change the default folders that iCloud uses:

- **Download folder** The My Photo Stream folder in your Pictures\Photo Stream\ folder—for example, C:\Users\Chris\Pictures\Photo Stream\My Photo Stream\ if your user account is named Chris
- **Upload folder** The Uploads folder in your Pictures\Photo Stream\ folder—for example, C:\Users\Chris\Pictures\Photo Stream\Uploads\ if your user account is named Chris

To set up Photo Stream on your PC, follow these steps:

1. If you don't already have iTunes on your PC, download the latest version from www.apple.com/itunes/ and install it. You must be running iTunes 10.5 or a later version to use iCloud and Photo Stream.
2. Chose Start | All Programs | Apple Software Update to run the Apple Software Update program, which checks for updated versions of iTunes and new components you need. Figure 2-8 shows Apple Software Update ready to download and install updates.

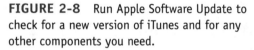

FIGURE 2-8   Run Apple Software Update to check for a new version of iTunes and for any other components you need.

3. Select the check box for each item you need to install. For example, in Figure 2-8, I've selected the check box for a new version of iTunes, the check box for a new version of QuickTime, and the check box for the iCloud Control Panel. I've refused the Safari 5 web browser.

    QuickTime is an Apple program that iTunes uses for playing back audio and video. To use all the features iTunes offers, you must install QuickTime on your computer. So if Apple Software Update offers you a new version of QuickTime, download and install it.

4. Click the Install Items button to download and install the items you've chosen. You may need to accept one or more end user license agreements to proceed.
5. Restart your PC if Apple Software Update prompts you to do so (as shown here), and then log back in.

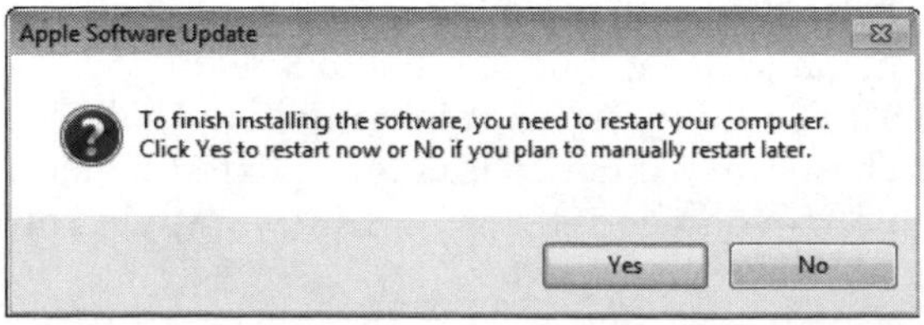

6. Click the Start button to open the Start menu.
7. Type **icloud** in the Search box, and then click the iCloud result that appears. The iCloud sign-in dialog box opens, as shown here.

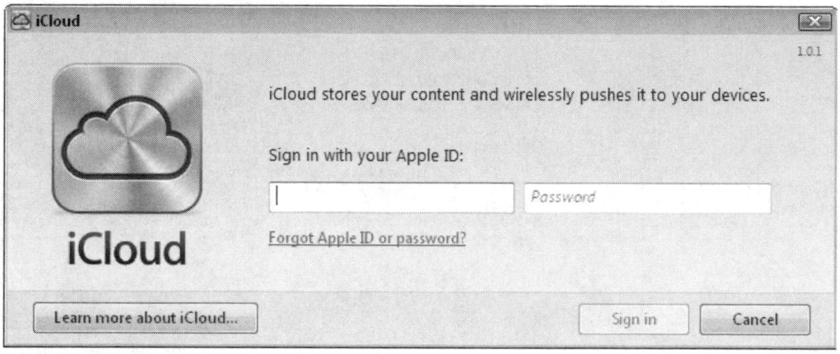

8. Type your Apple ID in the Sign In With Your Apple ID text box and your password in the Password text box.
9. Click the Sign In button. The iCloud dialog box shown in Figure 2-9 appears.
10. Select the Photo Stream check box to turn on Photo Stream.

FIGURE 2-9 In this iCloud dialog box, select the check box for each iCloud feature you want to use.

11. If you want to verify or change the default Download folder or Upload folder, click the Options button to the right of the Photo Stream check box to display the Photo Stream Options dialog box (shown here).

12. Click the Change button on the Download Folder line, select the folder you want in the Browse For Folder dialog box, and then click the OK button.
13. Click the Change button on the Upload Folder line, select the folder you want in the Browse For Folder dialog box, and then click the OK button.
14. Click the OK button to close the Photo Stream Options dialog box and return to the iCloud dialog box.
15. Click the Apply button to apply your changes.
16. Click the Close button to close the iCloud dialog box. The Close button appears in place of the Cancel button when you click the Apply button.

Now make sure that Photo Stream is working. Follow these steps:

1. Choose Start | Pictures to open a Windows Explorer window showing your Pictures folder.
2. Double-click the Photo Stream folder to open it.
3. Double-click the My Photo Stream folder to open it.
4. Check that the photos from your Photo Stream appear in the folder.
5. Add to the Uploads folder any photos that you want to upload to your Photo Stream.

Set Up Photo Stream on Your Mac

To set up Photo Stream on your Mac, follow these steps:

1. Choose Apple | System Preferences to display the System Preferences window.
2. In the Internet & Wireless section, click the Mail, Contacts & Calendars icon to display the Mail, Contacts & Calendars screen (shown in Figure 2-10 with an iCloud account selected).
3. In the accounts list on the left, click your iCloud account to display its controls.

FIGURE 2-10 To turn on Photo Stream on your Mac, select the Photo Stream check box in the iCloud pane of the Mail, Contacts & Calendars screen in System Preferences.

If you haven't yet set up your iCloud account on your Mac, click the Add Account button in the left column of the Mail, Contacts & Calendars screen. Then click the iCloud button to display the iCloud dialog box, type your Apple ID and password, and then click the Sign In button. In the Automatically Set Up iCloud dialog box that opens, click the OK button if you want to use the automatic setup process; click the Manual Setup button if you want to make all the choices yourself. Your iCloud account then appears in the accounts list on the left on the Mail, Contacts & Calendars screen.

4. Select the Photo Stream check box.
5. Choose System Preferences | Quit System Preferences or press ⌘-Q to quit System Preferences.

Now that you've set your Mac to use Photo Stream, it automatically downloads the photos that are currently in your Photo Stream. To see the photos, launch iPhoto, click the Photo Stream item in the Recent category in the Source list, and then click the Turn On Photo Stream button.

When you import photos from your camera or from an SD card into your iPhoto library, iPhoto automatically uploads the photos to Photo Stream, so they appear on your iOS devices and other computers that use Photo Stream.

Get Your Mac Ready for iCloud

Your Mac must be running Lion in order to make the most of iCloud. Earlier versions, including Snow Leopard (Mac OS X 10.6), can't use all iCloud's features.

To get your Mac ready for iCloud, first make sure your Mac is running Mac OS X Lion 10.7.2 or a later version. The easiest way to check is to choose Apple | About This Mac, and then look at the Version readout in the About This Mac dialog box. If your Mac has an earlier version of Lion, click the Software Update button in the About This Mac dialog box, and then follow the prompts to download and install the latest updates.

Second, update iTunes to the latest version. If you just updated Lion and accepted all updates offered, you've already updated iTunes. If not, choose Apple | Software Update to run Software Update, and then install any iTunes update offered, together with any other updates your Mac would benefit from. (Usually, it's a good idea to install all the updates.)

Once you've made these updates, you can set up your iCloud account in the iCloud pane of the Mail, Contacts & Calendars screen in System Preferences.

To add other photos to your Photo Stream, select the photos in iPhoto, click the Share button in the lower-right corner of the iPhoto window, and then click Photo Stream on the pop-up panel.

Project 14: Take Macro, Telephoto, and Panorama Photos with Your iPhone's Camera

Your iPhone's standard lens takes pretty amazing photos considering how tiny the lens itself is and how small the sensor it uses is. In fact, the photos are good enough that you probably won't want to carry around a digital camera when your iPhone is always at hand and can do the job instead.

But while the iPhone's photos are very good, you'll probably want them to be great. In this project, we'll look at two ways of taking even better photos with your iPhone:

- Adding lenses to take macro, fisheye, and telephoto shots
- Unlocking the iPhone's hidden Panorama feature so that you can take panoramas

Increase the Camera's Power by Adding Lenses

Where the iPhone falls down compared to a dedicated digital camera is that its lens is fixed rather than providing macro features and zoom features. But you can get around this limitation by adding lenses to your iPhone.

Digital cameras use two kinds of zoom: optical zoom and digital zoom. Optical zoom involves moving one or more lenses to create the zoom effect and keeps the image captured at full quality. Digital zoom works by increasing the size of the pixels in the zoomed area, which produces lower-quality results. Your iPhone uses digital zoom rather than optical zoom.

Because the iPhone has no lens mount, the lenses need to be held in position in different ways:

- Most of the lenses you'll find are built into cases, which means replacing your current case (if you use one).
- Other lenses clip onto the iPhone—which also tends to mean the iPhone can't be wearing a case.
- Some lenses attach via a magnetic ring. This works well only for light lenses, and again means the iPhone can't have a case on.

You can find a wide variety of moderate-cost lenses on Amazon, eBay, and other sites. Amazon tends to be the better place to buy such lenses because the reviews give you an idea of the product's quality and usefulness.

For more-expensive and higher-quality lenses, you're usually better off looking at a specialist store such as Photojojo (http://photojojo.com/store/). Here you can find items such as these:

- **iPhone SLR Mount**   The iPhone SLR Mount (see Figure 2-11) is a case that enables you to mount Canon EOS lenses or Nikon SLR lenses to your iPhone. (There's a Canon model and a Nikon model.) The iPhone SLR Mount costs $249.
- **iPhone Lens Dial**   The iPhone Lens Dial (see Figure 2-12) is an aluminum jacket with three lenses—telephoto, wide angle, and fisheye—built in. The lenses are on a rotating dial that you turn to switch from one lens to another. The iPhone Lens Dial costs $249 and has two tripod mounts built in—one for portrait orientation, the other for landscape orientation.

FIGURE 2-11   The iPhone SLR Mount enables you to fit a full-on camera lens to your iPhone. (Photo courtesy of Photojojo, http://photojojo.com.)

FIGURE 2-12   The iPhone Lens Dial lets you switch instantly among fisheye, wide angle, and telephoto lenses. (Photo courtesy of Photojojo, http://photojojo.com.)

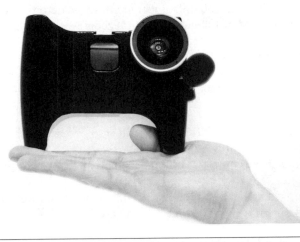

FIGURE 2-13 The iPhone Video Rig provides extra heft, an adjustable mic, two grips, and four tripod sockets. (Photo courtesy of Photojojo, http://photojojo.com.)

- **iPhone Video Rig**   The iPhone Video Rig (see Figure 2-13) is a chunk of milled aluminum designed to turn your iPhone into an easy-to-use video camera. Extra weight and two grips help you hold the iPhone steady, and four tripod sockets help you mount it firmly on a tripod. The iPhone Video Rig ($169) also has an adjustable mic so that you can avoid blocking the iPhone's built-in mic with your fingers.

Turn On Your iPhone's Secret Panorama Mode

To take panorama photos with your iPhone, you need to turn on Panorama mode. At this writing, Panorama mode is built into iOS 5, but it's not exposed to the user, presumably because Apple is still working on it. So you have to do a bit of hacking to enable Panorama mode on your iPhone.

Here's what we'll do in this section:

- Back up your iPhone so that you can restore your version of normality if things go wrong.
- Download and install an application called iBackupBot, then use it to edit one of the iPhone's configuration files.
- Restore your iPhone so that it uses the edited configuration file.
- Take panorama photos.

Back Up Your iPhone

First, back up your iPhone. Follow these steps:

1. Connect your iPhone to your PC or Mac. You can use Wi-Fi sync if you want, but using the USB cable is quicker.
2. If iTunes launches automatically and starts syncing your iPhone, let it finish. If you've configured iTunes not to launch and sync automatically, open iTunes yourself by clicking the iTunes icon on the Start menu or Taskbar (on Windows) or the Dock (on the Mac).
3. Click the iPhone's entry in the Source list to display its control screens.
4. If iTunes doesn't display the Summary screen at first, click the Summary button on the bar at the top to display it.
5. In the Backup box, make sure the Back Up To This Computer check box is selected.
6. Also in the Backup box, make sure the Encrypt iPhone Backup check box is not selected.
7. Right-click (or CTRL-click on the Mac) your iPhone in the Devices list and then click Back Up on the context menu, as shown here.

8. When the backup finishes, verify that the Last Backed Up To This Computer readout at the bottom of the Backup box shows Today and the time the backup finished.

Leave your iPhone connected to your computer for the moment, preferably with the Summary screen still displayed.

Download and Install iBackupBot and Edit the Preference File

Next, download and install iBackupBot. Follow these steps:

1. Open your web browser and go to the Download page on the iCopyBot website, www.icopybot.com/download.htm.

2. Download the free trial of iBackupBot for Windows or for the Mac and install it on your computer. The trial version is limited but allows you to edit the iPhone configuration files in the way needed here.

3. Launch iBackupBot from the Start menu (on Windows) or Launchpad or the Applications folder (on the Mac).

4. In the left pane, expand the iTunes Backups list, and then click the backup you want to use—normally your latest backup. The right pane displays the list of preference files contained in the backup, as shown in Figure 2-14.

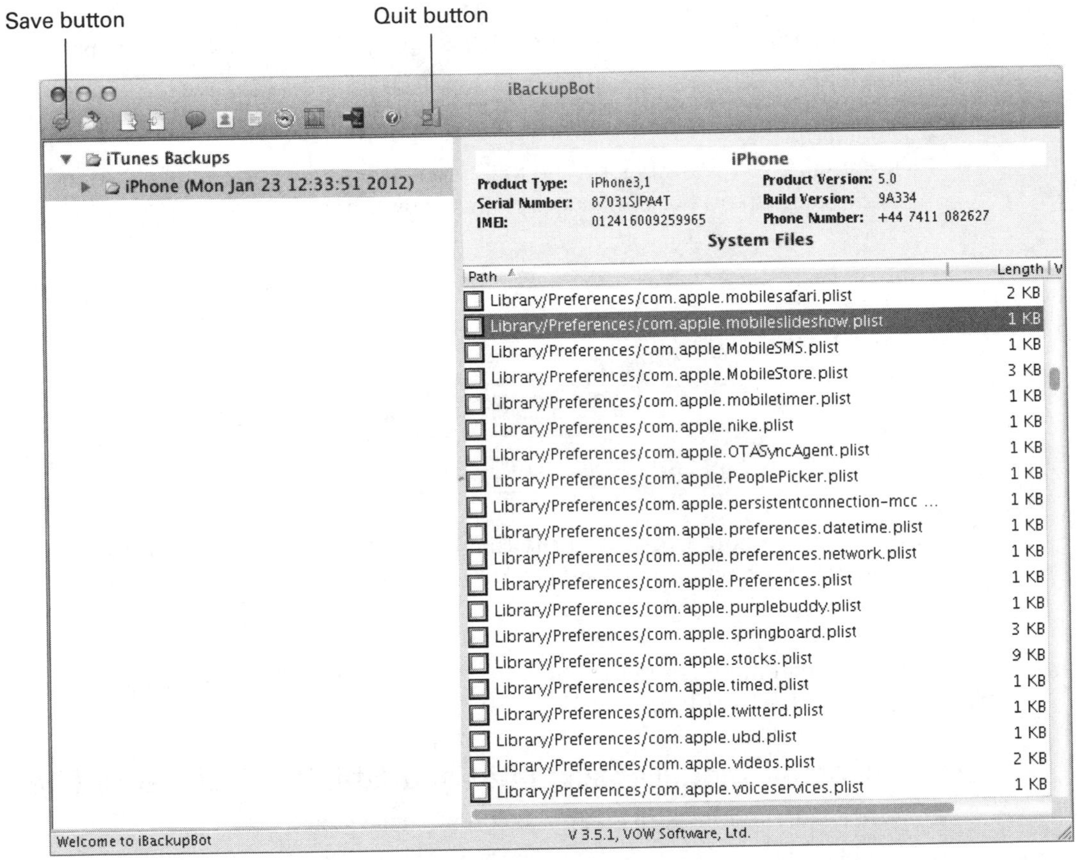

FIGURE 2-14   In the left pane, click the iPhone backup you want to use, then double-click the com.apple.mobileslideshow.plist file in the right pane.

5. In the right pane, double-click the com.apple.mobileslideshow.plist file to open it in the Editor window (shown here).

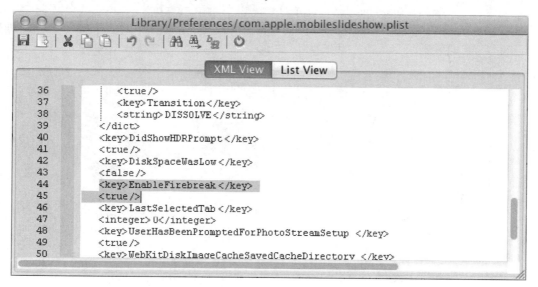

```
                    Library/Preferences/com.apple.mobileslideshow.plist

                              XML View    List View

 36               <true/>
 37               <key>Transition</key>
 38               <string>DISSOLVE</string>
 39       </dict>
 40       <key>DidShowHDRPrompt</key>
 41       <true/>
 42       <key>DiskSpaceWasLow</key>
 43       <false/>
 44       <key>EnableFirebreak</key>
 45       <true/>
 46       <key>LastSelectedTab</key>
 47       <integer>0</integer>
 48       <key>UserHasBeenPromptedForPhotoStreamSetup </key>
 49       <true/>
 50       <key>WebKitDiskImageCacheSavedCacheDirectory </key>
```

6. Find the pair of lines that reads like this:

```
<key>DiskSpaceWasLow</key>
<false/>
```

7. Click to place the insertion point after <false/> on the second line, and then press ENTER or RETURN to create a new line.

8. Type the EnableFirebreak key and its value, as shown here:

```
<key>EnableFirebreak</key>
<true/>
```

9. Click the Save button at the left end of the toolbar to save the change.
10. Click the Quit button at the right end of the toolbar to close the Editor window.
11. Close iBackupBot. For example, choose File | Exit (on Windows) or iBackupBot | Quit iBackupBot (on the Mac).

Restore Your iPhone with the Updated Configuration File

Now go back to iTunes and restore your iPhone. Follow these steps:

1. Display the Summary screen of the iPhone's control screens in iTunes if it's not already displayed.

2. Right-click (or CTRL-click on the Mac) the iPhone's entry in the Devices category in the Source list, and then click Restore From Backup on the context menu. iTunes displays the Restore From Backup dialog box (shown here).

3. Make sure the iPhone Name drop-down list shows the iPhone's name. (This shouldn't be a problem unless you sync multiple iPhones with this computer—but check anyway.)
4. Click the Restore button. iTunes restores the iPhone from the backup you created, and in doing so loads the preference file you edited.
5. When the restore operation finishes, disconnect your iPhone.

Take Panorama Photos on Your iPhone

You can now take panorama photos on your iPhone. Follow these steps:

1. Open the Camera app by tapping its icon on the Home screen.
2. Tap the Options button to display the camera options, as shown here.
3. Tap the Panorama button to turn on the Panorama feature. The Camera app displays instructions, as shown on the left in Figure 2-15.
4. Tap the Camera button to start taking the panorama photo.
5. Turn to your right, keeping the white arrow on the white line at the top of the screen (as shown on the right in Figure 2-15).
6. When you reach the end of the panorama, tap the Camera button to stop taking the panorama.

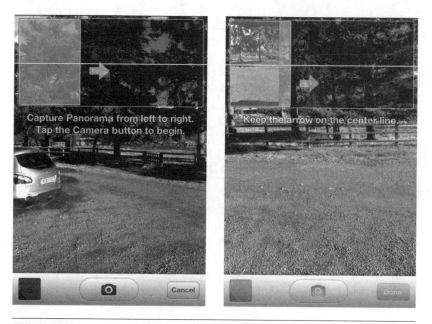

FIGURE 2-15   After turning on the Panorama feature, tap the Camera button to start taking a panorama photo. Turn to your right, keeping the white arrow on the white line (right).

Project 15: Take High-Quality Self-Portraits

Your iPhone's built-in front-facing camera—the camera on the screen side—is fine for FaceTime, and it works for quick self-portraits. Because you can see yourself on the screen, you can check the composition of the shot and make sure you adopt exactly the right smile or grimace before you release the shutter. But you're limited to the front-facing camera's lower resolution—640 × 480 pixels, also called VGA resolution.

So if you want to take high-quality self-portraits, you need to use the rear-facing camera, which has much higher resolution—3264 × 2448 pixels on the iPhone 4S (giving 8-megapixel photos), and 2592 × 1936 pixels on the iPhone 4 (giving 5-megapixel photos). Not only do you get better picture quality, but the rear-facing camera has a flash, whereas the front-facing camera has to make do with the available light.

 The easiest way to get high-quality photos of yourself using your iPhone's rear-facing camera is to have someone else take the photos for you. You don't need me to tell you how to do this. But other times you may want to control the entire process. In this project, I'll show you how to do that.

To take photos of yourself using your iPhone's rear-facing camera, you need to do two things:

- Mount the iPhone on a tripod.
- Get an app that gives you a self-timer function.

Let's look at these in turn.

Mount Your iPhone on a Tripod

First, you need to mount your iPhone on a tripod.

You can get tripods specially designed for the iPhone. In fact, if you put the search term **iPhone tripod** into eBay, you'll get more results than you could shake a large stick at. Amazon has plenty too. And if you look through specialist photography stores online, you'll find lots more.

As usual, the more you pay, the better built the tripod is likely to be. For example, a more expensive tripod may be made of aluminum rather than plastic. But because the iPhone is so light, even a flimsy plastic tripod is enough to keep it stable as long as conditions aren't too windy. So you don't need to break the bank to buy a heavy-duty tripod.

Figure 2-16 shows a lightweight table-top tripod that does a good job in domestic situations. As you can see, the tripod has a spring-loaded pressure clip to hold the iPhone steady.

 If you buy a tripod specially designed for the iPhone, make sure that the iPhone holder is detachable and has a regular tripod thread so that you can mount it on a full-size tripod as well.

Unless you want to shoot worm's-eye views (or you carry a table with you), a table-top tripod won't do you much good outdoors. What you'll probably want is a regular tripod that works for any camera that has a screw thread. Chances are you've got such a tripod already somewhere in your home; if not, you can beg, borrow, or buy one easily enough.

To mount your iPhone on the tripod, you need a bracket that grips the iPhone and provides a threaded hole for attaching it to

FIGURE 2-16 A tripod designed for the iPhone usually has a spring-loaded pressure clip to hold the device steady.

the tripod's screw. If you've bought a table-top tripod, see if you can take the top off that and mount it on a regular tripod. If not, buy a tripod bracket separately.

Figure 2-17 shows the iPhone mounted on a full-size tripod using the pressure clip from the table-top tripod. The arrangement looks a bit funny, especially from the subject's side, because the iPhone doesn't have the hefty lens you might expect to see peering at you. But it works well, which is all we care about here.

Searching for **iPhone tripod mount** or **iPhone tripod bracket** on eBay or Amazon will find you plenty. A plastic bracket will do the job, but if you have the choice, go for one that has a metal hole rather than a plastic hole for attaching the bracket to the tripod's screw. The plastic holes tend to wear out quickly, especially if you over-tighten them.

FIGURE 2-17   Mounting your iPhone on a full-scale tripod enables you to take high-quality self-portraits and other shots that require stabilization.

Get an App with a Self-Timer Function

What you need next is a camera app with a self-timer function. The iPhone's built-in Camera app doesn't have a self-timer, so you need to go for a third-party app.

If you've already bought a third-party camera app, check to see whether it has a self-timer. If not, you have various choices, but your best free choice at this writing is Gorillacam.

Gorillacam is a camera-enhancement app created by Joby, the company that makes the Gorillapod family of flexible tripods. Apart from its price being right, Gorillacam has a handful of great features, including the self-timer (which is what we want here) to a time-lapse capability that's great for capturing sunrises—or for surveillance, if that's what you need. You can set the self-timer for anything from 3 seconds up to 90 seconds—enough time not only to get yourself in the shot but to brush your hair as well.

To download and install Gorillacam, follow these steps:

1. Activate the iTunes window.
2. Double-click the iTunes Store item in the Source list on the left to open a window showing the iTunes Store.
3. Type **gorillacam** into the Search box and press ENTER or RETURN.

4. Click the appropriate search result to display the app's page.
5. Click the button to buy the app, and then confirm the purchase.

You can also get Gorillacam by using the iTunes Store on your iPhone if you prefer.

After downloading and installing Gorillacam, you can quickly set up the self-timer like this:

1. Tap the Settings button in the lower-left corner of the Gorillacam screen to display the Settings screen (shown on the left in Figure 2-18).
2. Tap the Self-Timer switch and move it to the On position. Gorillacam displays the Self-Timer button between the Settings button and the Shutter Release button at the bottom of the screen.
3. Tap the Settings button again to close the Settings screen.
4. Tap the Self-Timer button to display the Delay pop-up menu (shown on the right in Figure 2-18).
5. Tap the delay you want—for example, 10 Secs.

Now when you're ready to take the photo, tap the Shutter Release button, and the countdown starts.

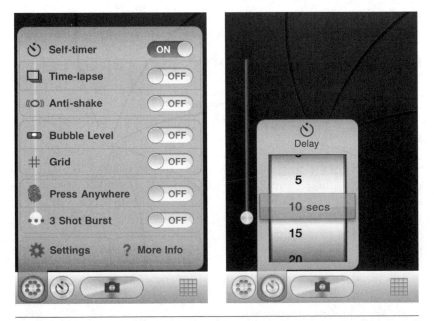

FIGURE 2-18 Tap the Settings button in the lower-left corner of the Gorillacam screen to display the Settings screen (left), then tap the Self-Timer switch and move it to the On position. You can then tap the Self-Timer button and set the delay.

Project 16: Take Time-Lapse Movies and Shoot Video at Different Frame Rates

Your iPhone's camera takes high-definition video at 30 frames per second:

- **iPhone 4S**   1920 × 1080 pixels (1080p)
- **iPhone 4**   1280 × 720 pixels (720p)

This quality is high enough for capturing broadcast-quality video, so you'll want to make the most of it—for example, posting your edited video clips on YouTube or making them into movies using iMovie either right there on your iPhone or on your Mac.

But if you're serious about capturing video and making movies, you'll probably want to go beyond what the Camera app can do. You can do so by installing a third-party app that gives you control over how the camera shoots video: choosing the resolution, setting the frame rate, locking the focus or exposure, and so on.

In this project, we'll look at how you can change the frame rate to make time-lapse movies and to shoot footage that will appear to run at higher speed. For example, if you shoot video at 15 frames per second but play it back at normal speed, everything will seem to happen twice as fast. And if you take a time-lapse video of a sunrise at a single frame per second, each minute of real time will be compressed into two seconds of video when you play it back.

Get FiLMiC Pro

At this writing, the best app for adding capabilities to your iPhone's camera is FiLMiC Pro, which costs $3.99. FiLMiC Pro gives you control over the camera's frame rate, exposure, white balance, resolution, and other settings.

Your first move is to get FiLMiC Pro. Follow these steps:

1. Activate the iTunes window.
2. Double-click the iTunes Store item in the Source list on the left to open a window showing the iTunes Store.
3. Type **filmic pro** into the Search box and press ENTER or RETURN.
4. Click the appropriate search result to display the app's page.
5. Click the button to buy the app, and then confirm the purchase.

 You can also get FiLMiC Pro by using the iTunes Store on your iPhone if you prefer.

After iTunes downloads the app, sync your iPhone to install it. Depending on your sync settings, you may need to select the app's check box on the Apps screen in iTunes to install it on the iPhone.

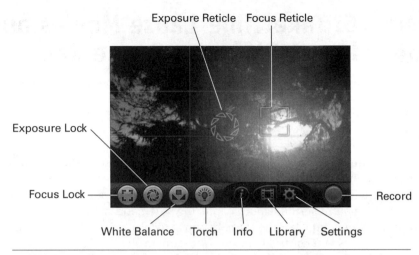

Exposure Reticle   Focus Reticle

Exposure Lock

Focus Lock

Record

White Balance   Torch   Info   Library   Settings

FIGURE 2-19 FiLMiC Pro provides separate reticles for focusing and exposure. You can also adjust the white balance and the frame rate.

Launch FiLMiC Pro

After installing FiLMiC Pro, tap its icon on the Home screen to run it. FiLMiC Pro displays whatever the camera is seeing, as shown in Figure 2-19.

The interface is pretty straightforward to use. For example, you tap the focus reticle and drag it to the area of the screen where you want to focus; similarly, you tap the exposure reticle and move it to the area of the screen on which to meter the exposure.

Adjust the Frame Rate

To adjust the frame rate, follow these steps:

1. Tap the Settings button on the FiLMiC Pro screen to display the Settings screen (shown on the left in Figure 2-20).
2. Tap the FPS button to display the Frame Rate screen (shown on the right in Figure 2-20).
3. Tap the frame rate you want. Your choices range from the iPhone's top speed, 30 fps, all the way down to a single frame per second, 1 fps.
4. Tap the Settings button to return to the Settings screen.
5. Tap the Done button to return to the camera.

Shoot Your Video

With the frame rate set the way you want it, you're now ready to shoot your video. Mount your iPhone on a tripod as discussed in the previous project, line up your subject, and then tap the Record button.

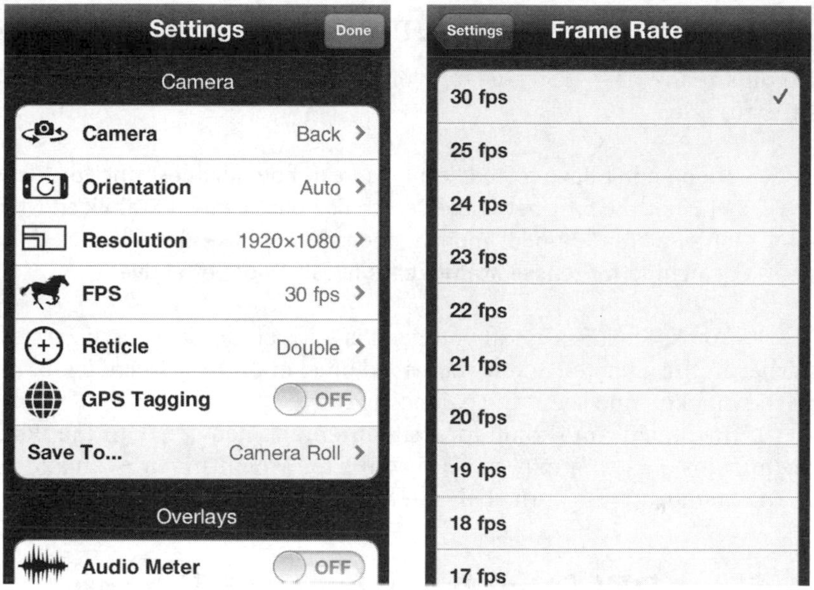

FIGURE 2-20   On the Settings screen (left), tap the FPS button to display the Frame Rate screen (right), and then tap the frame rate you want to use.

Project 17: Adapt or Build a Steadicam to Stabilize Your iPhone for Shooting Video

Your iPhone shoots high-definition video, and it gives great results for its size. But it has a problem when your subject matter is moving.

Whereas a full-scale video camera uses a mechanical shutter to create separate video frames, your iPhone uses a *rolling shutter*, which takes several milliseconds to create each frame. The rolling shutter is not good at capturing movement, because a fast-moving subject can move while a frame is being captured. This results in blurred video.

Blur also occurs when you (the cameraperson) move the iPhone while capturing video. When you're shooting from a stationary position, you can keep the iPhone still by using a tripod, as described earlier in this chapter. But when you're moving, you need to use a device to stabilize the iPhone and damp down your movements so that the video you shoot appears smoother.

Such video-camera stabilization devices are usually called Steadicams. You can buy iPhone-specific Steadicam rigs, but if you already have a Steadicam, it makes more sense to adapt it, as described in the first part of this project. And if you don't have a Steadicam, you may prefer to build one rather than buy one. In the second part of this project, I'll show you how to build a DIY Steadicam from scratch.

Mount Your iPhone on an Existing Steadicam

If you already have a Steadicam, you should be able to mount your iPhone on it with little trouble.

 If you don't have a Steadicam, you can buy a modest one for $100 or less. At this writing, the best bet is the Lensse Compact Camera Stabilizer, which you can find on eBay for $70-odd. You may see this stabilizer described as the iSteady, but searching for **lensse stabilizer** is usually most effective.

For most Steadicams, all you need is a tripod bracket, as discussed in Project 15, earlier in this chapter. Screw the tripod bracket onto the Steadicam, mount your iPhone in the bracket, and see if the balance is okay.

If the Steadicam is built for a camera much heavier than the iPhone (as most Steadicams are), you may need to adjust the weighting or even add weight at the top of the Steadicam to get the balance right.

Build a DIY Steadicam for Your iPhone

If you don't have a Steadicam, and you don't want to buy one, you can build yourself one inside a couple of hours largely from items you probably have around your home or yard—although there's one thing you'll probably need to buy.

This section shows you how to build a Steadicam from an old bike wheel.

 This Steadicam is inspired by one created by Thomas Johnson. To see Johnson's Steadicam, go to YouTube and search for **thomasumjohnson**.

Get the Things You Need for the Steadicam

Here's what you'll need to build the Steadicam:

- **Bike wheel**   A wheel from a kid's bike is best—for example, a 16- or 18-inch wheel. You can use a full-size (26- or 27-inch) bike wheel if you want, but the result is bigger than you'll probably want unless you need to be able to mount larger cameras rather than just your iPhone.
- **Tripod bracket**   You need a tripod bracket to mount your iPhone on the top of the Steadicam. The Steadicam has a standard tripod screw at the top, so you can use the same tripod bracket as for attaching the iPhone to your tripod.
- **Tripod head**   To enable you to point the iPhone at the angle you want, you'll need a tripod head.
- **Gimbal or universal joint**   To damp your movement, the Steadicam needs a joint that can turn freely in two directions. The best thing is a brass gimbal made by Lensse. You can pick these up on eBay from $15 upward. Alternatively, you can use a universal joint such as those made by Traxxas for its Slash radio-controlled cars.

A *gimbal* is a device that can turn freely in two or three directions to keep an instrument level. Most gimbal designs involve several rings that pivot at right angles to each other.

- **Counterweight**   To get the balance right, the Steadicam needs a counterweight at the bottom. This can be very basic—for example, a piece of scrap metal. I used a small weight plate from a dumbbell set. The weight you need will depend on the other items you use for the Steadicam, but typically it'll be in the range of one pound to several pounds.
- **Tools**   You'll need a modest number of bike-repair and metalworking tools:
  - Hacksaw
  - Metal file
  - Tire levers
  - Spoke wrench
  - Standard wrenches
  - Screwdriver (bonus points for electric)
  - Reamer

Build the Steadicam

To build the Steadicam, follow these steps:

1. If the bike wheel has a tire on it, take the tire off.
2. Unscrew the axel and take it out. Keep it—you'll use it later.

 DOUBLE GEEKERY

Build Your Own Gimbal from Scrap or Pipe

The gimbal is the most expensive part of the DIY Steadicam, and you may balk at the price—especially given that you can buy an inexpensive Steadicam (which includes a gimbal) instead for three or four times as much.

If so, you can make your own gimbal using a design such as that shown here.

As you can see, what you need are three metal or plastic rings of sizes that will fit inside each other, and bolts to secure them. Because you won't put serious weight on the gimbal, the rings can be of lightweight material—for example, jar lids or plastic pipes. Mount the smallest ring on the middle ring with two bolts north and south so that it can turn freely, and then mount the middle ring on the larger ring with two bolts east and west so that it can turn freely in a different dimension.

3. Remove the axel housing and keep that too.
4. Remove the protective strip that covers the ends of the spokes where they connect to the rim of the wheel.
5. Undo each spoke and remove the hub.

Undo each spoke by turning its fastener with the spoke wrench until the spoke itself is below the screw slot at the top of the fastener. Then use the screwdriver to unscrew the fastener. An electric screwdriver will save you time and effort here. To prevent the hub from hanging, leave a spoke attached at each side—north, south, east, and west, as it were—until you've removed all the other spokes.

6. Saw the rim into two pieces:

Look at the join in the wheel rim and see if you can pull it apart. Some rims are easy to separate, but others are joined so tightly that sawing the rim is easier.

- The piece you'll use for the Steadicam needs to be more than half of the circle, so that there's some overlap at the top for mounting the camera and at the bottom for mounting the counterweight.
- Usually, you'll want at least 210° of the 360° in a circle—enough to make a C shape, as shown here.
- You can either eyeball the measurement—it doesn't have to be exact—or count spoke holes. If you're counting spoke holes, divide the total number of spoke holes by 0.58 (the decimal for 21/36, or 210/360). For example, if the wheel has 48 spokes, you'll want 28 spoke holes on the longer section. If the wheel has 36 spokes, you'll want 21 spoke holes on the larger section. If the wheel has 24 spokes, you'll want 14 spoke holes on the larger section.

In many rims, the spoke holes are off center on alternate sides. This is fine for attaching the handle and the counterweight, but you'll probably find it easier to attach the tripod head to the valve hole, which is centered in the rim. So when you cut, have the valve hole just before the cut at the top of the C shape.

- If in doubt, cut a longer section than you need. You can easily cut it shorter.
7. File down the edges of the cut rim.
8. If necessary, use a metal reamer to ream out the valve hole to enable the bolt securing the tripod head to go through it.

9. Attach the tripod head to the valve hole with a bolt, as shown here.
10. Ream out a spoke hole a little way back from the tripod head at the top of the C so that the end of the axel will pass through it.
11. Attach one end of the axel to the hole in the rim, so that the main part of the axel is on the inside of the C, as shown below.

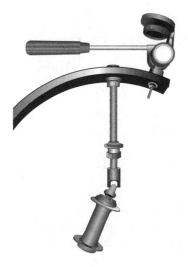

12. Attach the gimbal or universal joint to the free end of the axel.
13. Attach the hub (or whatever you're using as the handle) to the gimbal or universal joint, as shown in this illustration.

14. Attach the counterweight to the bottom of the C shape, as shown in the illustration. If you're using a weight plate, mount it flat so that it will act as a crude stand for the Steadicam.

Now mount the iPhone in the tripod bracket and screw it onto the top of the tripod head. You've now got a Steadicam rig that'll stabilize the video you shoot on your iPhone while you're moving. You're ready to start making movies on location.

Project 18: View Your Webcam on Your iPhone

If you use a webcam to keep tabs on what's happening at home when you're away, you can tap into the webcam from your iPhone to stay on top of things no matter where you happen to be.

You can also turn your iPhone into a network webcam that you can monitor from any other computer on your network. We'll look at how to do this toward the end of this section.

Decide Which Software to Get

To view your PC's or Mac's webcam on your iPhone, you need two software applications:

- **PC or Mac application**   On your PC or Mac, you run an application that will shunt the video signal out across your local network or the Internet to where your iPhone can receive it.
- **iPhone app**   On your iPhone, you run an app that can connect to the video stream your PC or Mac is providing, and display the pictures to you.

In this section, we'll look at an iPhone app that has companion software for both Windows and the Mac: Air Cam Live Video. The full version of Air Cam Live Video costs $7.99, but there's a free version called Air Cam Live Video (Lite) that you'll probably want to try first.

If you need to monitor only a webcam connected to a Windows PC, look also at JumiCam. Start with JumiCam Lite, which is free but limited to your local network, and see how well it works for your needs. If you need extra features, such as reaching across the Internet to monitor your webcam, upgrade to JumiCam Pro ($7.99).

Get and Set Up the Software on Your PC or Mac

First, download and install the desktop software needed: Air Cam Live Video for Windows or Air Cam for the Mac.

Download the Software for Your PC or Mac

Open your web browser and go to the Air Cam Live Video for iOS page on the Senstic website (www.senstic.com/iphone/aircam/aircam.aspx). Then click the Download Air Cam Live Video For Windows XP/Vista/7 link or the Download Air Cam Live Video For Mac OS X link, as appropriate.

Install and Run the Air Cam Live Video for Windows Program

When downloading the AirCamSetup.msi file, choose the option to run it (or to save it and run it, depending on your browser). When the Air Cam Setup Wizard runs, follow through its prompts. You'll need to install some codecs (*coder/decoder* software) unless your PC already has them, so the installation process has multiple steps, and you need to make some decisions.

These are the key points in the installation:

- Close Internet Explorer before you run the installation.
- The Select Installation Folder screen lets you choose to install Air Cam in a different folder than the default folder (a Senstic\Air Cam\ folder inside your Program Files folder), but normally you're safe sticking with the default folder.
- On Windows 7 or Windows Vista, you will need to click the Yes button in the User Account Control dialog box (shown here) to continue with the installation. Make sure the User Account Control dialog box gives the program name Air Cam Installer.

- When the installer displays the Additional Packages dialog box (shown here), click the Get K-Lite button to open a browser window to a site that provides the K-Lite pack of codecs. Follow the links to download the K-Lite pack and then install it. At this point, the Air Cam Installer is still running, but it's in the background.

When downloading the K-Lite codec pack, be sure to click the correct link. The web page may contain tempting Download buttons for other software the web hosts would like you to try.

- When the Setup – K-Lite Codec Pack installer runs, you'll get another User Account Control dialog box, this time for the K-Lite Codec Pack. You'll need to click the Yes button in this dialog box to proceed with the installation.
- On the initial screen of the Setup – K-Lite Codec Pack installer (shown here), click the Simple Mode option button. Click the Next button and go through the next several configuration screens. You'll probably want to accept the default settings here.

- On the Additional Options screen of the Setup – K-Lite Codec Pack installer (shown here), select the "No Thanks. I Don't Want Any Of The Above" check box to prevent the installer from burdening your PC with the StartNow Toolbar, setting your home page to StartNow, and making Yahoo! your default search engine.

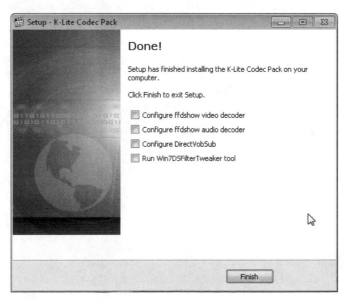

- When you reach the Ready To Install screen, click the Install button.
- When the Done! screen (shown here) appears, make sure all the check boxes are cleared, and then click the Finish button. The Setup – K-Lite Codec Pack installer closes.

- Back in the Additional Packages dialog box, click the Install Bonjour button to install Apple's Bonjour networking protocol. The Bonjour Print Services installer then starts and displays its Welcome screen (shown here). Click the Next button and go through the screens for accepting the license agreement and reading information about Bonjour.

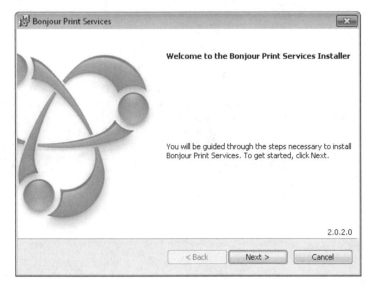

 If the Install Bonjour button in the Additional Packages dialog box is dimmed and unavailable, your PC already has Bonjour for Windows installed, so you don't need to install it. Go ahead and click the Exit button to close the Additional Packages dialog box.

- On the Installation Options screen (shown here), clear the Create Bonjour Printer Wizard Desktop Shortcut check box unless you want a shortcut for the Bonjour Printer Wizard on your desktop. Clear the Automatically Update Bonjour Print Services And Other Apple Software check box if you don't want the Apple Software Update service to check automatically for updates. (You may prefer to check manually for updates at times that suit you.) Then click the Install button.

- When the Congratulations screen of the Bonjour Print Services installer appears, click the Finish button.
- At this point, you'll see the Air Cam installer window again. Click the Close button, and you're finally done with the installation.

After installing Air Cam Live Video, run it by choosing Start | All Programs | Senstic | Air Cam | Air Cam Live Video.

If Windows displays a User Account Control dialog box, check that the program name is Air Cam for Windows, and then click the Yes button.

You can now enter your access information and configure the webcam as described in the section "Configure Air Cam Live Video or Air Cam," later in this project.

Install and Run the Air Cam Application on Mac OS X

After downloading the Air Cam application for Mac OS X, install it like this:

1. Click the Downloads icon on the Dock to display a stack showing your downloaded files, and then click the Air Cam.pkg.zip file. Mac OS X unzips the file and displays a Finder window showing the Downloads folder with the Air Cam.pkg file selected.
2. Double-click the Air Cam.pkg file to launch the Air Cam for Mac installer.
3. Click the Continue button to display the Installation Type screen. Here, you can click the Customize button to display the Customize screen if you want to prevent the installer from installing the AirCamLauncher app, but normally you'll be best off accepting the Standard install.

4. Click the Install button to run the installation, and then type your password (or an administrator's password) in the Authenticate dialog box.
5. When the installer shows the "The installation was successful" screen, click the Close button.
6. In the Finder window, click Applications in the sidebar to display the Applications folder.
7. Double-click the Air Cam icon to launch Air Cam.

Configure Air Cam Live Video or Air Cam

The first time you start Air Cam Live Video (on Windows) or Air Cam (on the Mac), the program displays the Enter Access Information dialog box. This dialog box (the Mac version appears in the next illustration) prompts you to enter an e-mail address and password for accessing Air Cam Live Video or Air Cam from another computer. This is what you'll need to access Air Cam Live Video or Air Cam from your iPhone.

```
                Enter Access Information

        Please enter a valid E-Mail address and a
        password to enable remote access of Air Cam:

        This information also allows your iPhone/iPod
            to connect without using computer's IP

        Access Information

        E-Mail:   [                        ]

        Password: [                        ]

                                        ( Set )
```

Type text that looks like an e-mail address in the E-Mail box—you don't need to enter a real e-mail address, and you may prefer not to enter a real address for security reasons. For example, enter *notmyname@example.com*.

Then click in the Password box and type the password you will use to connect your iPhone to Air Cam Live Video or Air Cam. Make this a strong password—at least six characters; including upper- and lowercase letters; including at least one number and at least one symbol; and not a real word in any language.

Click the Done button (on Windows) or the Set button (on the Mac) to close the Enter Access Information dialog box. You'll then see the Air Cam Live Video window (on Windows) or the Air Cam window (on the Mac). The left screen in Figure 2-21 shows the Air Cam Live Video window; the right screen in Figure 2-21 shows the Air Cam window.

Aim the webcam (or the computer, if the webcam is built in) so that the picture shows what you want to watch. Then click the Options button to display the Air Cam

FIGURE 2-21   The Air Cam Live Video window (left) shows what your PC's webcam is seeing. The Air Cam window (right) shows the webcam's view on your Mac.

Options dialog box. On the six screens in this dialog box, you can set up Air Cam Live Video to work the way you prefer:

- **Webcam**   This screen (shown here) shows options you'll want to set (or at least verify) to get Air Cam Live Video working correctly: You can set the resolution, flip the webcam's picture vertically or horizontally, turn on the display of the time stamp or the camera's name, and turn on Night Vision mode (for use in low light).

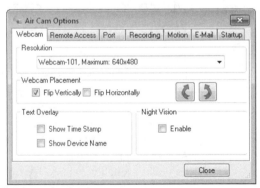

- **Remote Access**   On this screen, you can change the e-mail address and password you're using for remote access. You can also set up port forwarding by using the Auto-Config feature for Universal Plug and Play.

- **Port**   On this screen, you can change the default TCP listener port. The standard setting is port 1726.

- **Recording**   On this screen, you can select the folder to which Air Cam Live Video saves recorded video files. (You can start and stop recording from your iPhone once you connect.)

- **Motion**   On this screen, you can configure motion-detection settings. You can adjust the sensitivity level by clicking the Low option button, the Medium option button, or the High option button. You can choose what Air Cam Live Video should do when it detects movement: send you an e-mail, send you a push notification, or start recording automatically.
- **E-Mail**   On this screen, you can set up the e-mail account to which Air Cam Live Video should send notifications.
- **Startup**   On this screen (which appears only on Windows), you can choose to run Air Cam Live Video automatically at startup, and you can make Air Cam Live Video start up in hidden mode (so that you see only an icon in the notification area rather than a window showing what the webcam is seeing).

Click the Close button when you finish choosing options in the Air Cam Options dialog box. Air Cam Live Video is now running, and you can connect to it from your iPhone as described later in this section.

Get and Set Up Air Cam Live Video on Your iPhone

Now that you've got Air Cam Live Video running on your PC or Mac, your next move is to install Air Cam Live Video on your iPhone and set it up to connect to your PC or Mac.

DOUBLE GEEKERY

Make Your Internet Router Pass Requests from Your iPhone to Air Cam

To connect Air Cam on your iPhone to Air Cam Live Video (on Windows) or Air Cam (on the Mac) across the Internet, you need to set your Internet router to pass incoming requests for Air Cam to your PC or Mac. To do this, you set the router to forward traffic on the appropriate port. The default port is TCP port 1726.

If your Internet router supports the Universal Plug and Play (UPnP) standard, you can do this automatically by clicking the Auto-Config (UPnP) button on the Remote Access screen in the Air Cam Options dialog box. Many Internet routers support UPnP, so you'll probably want to try this approach first.

If this doesn't work, go into your router's configuration screens and make sure UPnP is turned on, and then try again. You may find that UPnP is turned off as a security measure. (Or you may have turned it off yourself.)

If your Internet router doesn't support UPnP, or if it does but you prefer to leave UPnP turned off, you can set up the port forwarding yourself. Go into your router's configuration screens, find the screen for port forwarding (or port redirection), and set up a rule to forward TCP port 1726 to your PC or Mac. For best results, you may need to give your PC or Mac a fixed IP address rather than have your Internet router allocate an address via DHCP.

Activate the iTunes window, and then double-click the iTunes Store item in the Source list on the left to open a window showing the iTunes Store. Type **air cam** into the Search box, press ENTER or RETURN, and then click the appropriate search result. On the app's page, click the button to download the Lite version for free or the Buy button to buy the full version.

You can also get Air Cam by using the iTunes Store on your iPhone if you prefer.

After iTunes downloads the app, sync your iPhone to install it. Depending on your sync settings, you may need to select the app's check box on the Apps screen in iTunes to install it on the iPhone.

After installing Air Cam, tap its icon on the Home screen to run it. If Air Cam displays the Please Turn Off Bluetooth dialog box (shown here), look at the right side of the status bar at the top of the screen to see if the Bluetooth icon appears. At this writing, Air Cam displays the Please Turn Off Bluetooth dialog box whether Bluetooth is on or off, not just when Bluetooth is on (as you'd expect).

If the Bluetooth icon appears on the status bar, follow these steps to turn off Bluetooth:

1. Press the Home button to display the Home screen.
2. Tap the Settings icon to display the Settings screen.
3. Scroll down to the third box.
4. Tap the General button to display the General screen.
5. Tap the Bluetooth button to display the Bluetooth screen.
6. Tap the Bluetooth switch and move it to the Off position.
7. Tap the General button to go back to the General screen.
8. Tap the Settings button to go back to the Settings screen.

Now press the Home button twice in quick succession to display the app-switching bar, and tap the Air Cam button on it to switch back to Air Cam. Tap the OK button in the Please Turn Off Bluetooth dialog box.

Connect to the Webcam from Your iPhone

Next, you see the Connection screen (shown on the left in Figure 2-22), which lists the webcams you can connect to.

Tap the webcam you want to view. If the webcam has a password (as is usually the case), Air Cam displays the Authentication screen (shown on the right in Figure 2-22) prompting you to enter it.

FIGURE 2-22 On the Connection screen (left), tap the webcam you want to connect to. If Air Cam displays the Authentication screen (right), type the password and tap the Go button.

Type the password, and then move the Remember It switch to the On position or the Off position, as appropriate. Saving the password will make future access to the webcam quicker, so you'll probably want to save it, unless doing so is too much of a security risk.

Tap the Go button at the lower-right corner of the keyboard to connect to the webcam. Its video appears on the screen, as shown in Figure 2-23. In portrait orientation, you can use the controls to change Air Cam's settings, adjust the picture, take snapshots and recordings, and sync audio and video:

- **Settings**   Tap this button to display the Settings screen, which gives you access to Air Cam's options.
- **Snapshot**   Tap this button to take a snapshot of the screen.
- **Remote Recording**   Tap this button to start making a recording of the video.
- **Source Selection**   Tap this button to change the video source.
- **Volume Control**   Tap this button to adjust the volume using the slider that appears on the right side of the screen.
- **Frame Rate**   Tap this button to adjust the frame rate using the slider that appears on the right side of the screen.
- **Audio/Video Sync Control**   Tap this button to adjust the audio/video sync using the slider that appears on the right side of the screen.

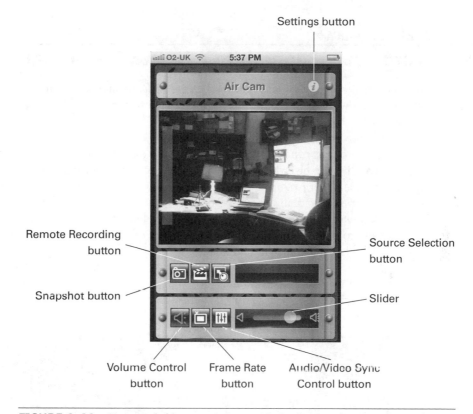

FIGURE 2-23   In portrait view, you can use Air Cam's buttons to configure the app itself and the video and audio it receives.

Air Cam is easy to control on your iPhone, but it appears to lack a command for shutting off the connection to a webcam you've been viewing. You can press the Home button to display the Home screen, but that leaves Air Cam running in the background. So to shut down the connection, and to shut down the app, you need to force quit Air Cam like this:

1. Press the Home button to display the Home screen.
2. Press the Home button twice in quick succession to display the app-switching bar.
3. Tap and hold the Air Cam button until a red circle containing a – sign appears at the upper-left corner of each app's icon in the app-switching bar.
4. Tap the – button for Air Cam to close the app.

DOUBLE GEEKERY

Turn Your iPhone into a Network Webcam

As you know, your iPhone has not one but two video cameras built in, together with a wireless network interface. So it's fully equipped to act as a network webcam itself if you need it to.

If you keep your iPhone with you at all times, you can use it to broadcast what you're doing so that others can watch it in their web browsers. If you don't mind parking your iPhone elsewhere for a while, you can use it to keep tabs on what's happening there. For example, you can set up your iPhone as a baby monitor, check on what your dogs actually do while you're out, or rig the iPhone to keep watch over your desk at work to see who's disappearing your donuts.

You can broadcast either on your local network or across the Internet, and tap into it with just about any web browser.

To use your iPhone as a network webcam, get the app ipCam – Mobile IP Camera ($2.99) from the App Store. Launch the app, and you'll see the Preview screen, which shows what the rear-facing camera is seeing (see the left screen in the illustration).

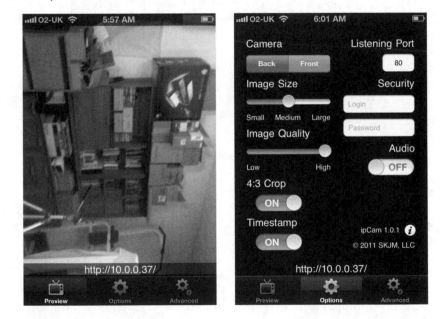

Next, tap the Options button at the bottom to display the Options screen (shown on the right in the illustration). Here you can set up the camera:

- **Camera**   Tap the Back button or the Front button to switch cameras.
- **Image Size**   Drag the slider along the Small–Medium–Large scale to set the image size you want. Large images work fine over a local network, but if you're monitoring across the Internet, you'll normally want to use the Medium size.
- **Image Quality**   Drag the slider along the Low–High scale to set the image quality.
- **4:3 Crop**   Move this switch to the On position if you want to crop the picture to a 4:3 aspect ratio. Set this switch to the Off position if you want to see the iPhone's picture straight.
- **Timestamp**   If you want to see the date and time superimposed on the iPhone's picture, set this switch to the On position.
- **Listening Port**   If necessary, change the number in this text box from the default (80) to a different port number. You may need to use a different port when accessing your iPhone across the Internet via a router.
- **Security**   If you want to secure the connection, type the login name in the Login box and the password in the Password box.
- **Audio**   Move this switch to the On position if you want to receive the iPhone's audio along with the pictures. Getting the audio increases the amount of data transmitted, but it can be a great help in working out what's going on.

When you finish choosing settings on the Options screen, tap the Advanced button at the bottom of the screen to display the Advanced screen. This contains three items:

- **External IP**   This readout shows the external IP address your network is using. This is the address at which you will be able to connect to the iPhone across the Internet from outside your network. You may need to change the settings on your Internet router to enable computers to access your iPhone across the Internet.
- **Dynamic DNS Update**   If your Internet provider uses dynamic IP addresses rather than static IP addresses, you can use either the DynDNS service or the No-IP service to provide the IP address when it changes. To use dynamic DNS, move this switch to the On position. In the controls that appear when the switch is in the On position, tap the DynDNS tab or the No-IP tab, as appropriate, and then fill in the fields.
- **Automatic Port Forwarding**   To use automatic port forwarding, move this switch to the On position. Automatic port forwarding lets ipCam tell your Internet router where to send incoming request for ipCam from the Internet. To use automatic port forwarding, your Internet router must support either Universal Plug and Play (UPnP) or NAT-PMP.

When you've set your iPhone up to monitor, open a web browser and go to the IP shown at the bottom of the ipCam window. You'll see the control screen, as shown on the left in the following illustration.

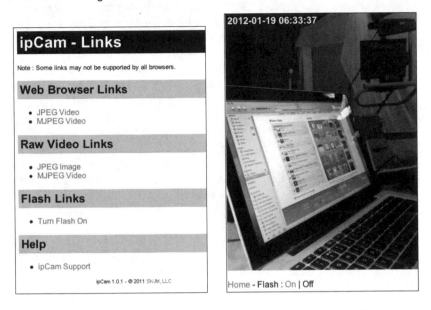

Click the JPEG Video link in the Web Browser Links area, and you will see a video feed from the iPhone, as shown on the right in the illustration. Click the MJPEG Video link in the Web Browser Links area, and you can get an audio feed along with the video feed. To use audio, you will need to have set the Audio switch on the Options screen to the On position.

If you find the picture is dark, click the Turn Flash On link in the Flash Links area to turn the flash on.

When using ipCam, you'll normally want to plug your iPhone into the USB Power Adapter to make sure you don't run through the battery. Providing power to the iPhone is doubly important if you plan to use the flash.

3 iPhone as Your Main Computer Geekery

Up till now, you've pretty much needed a full-size computer—PC or Mac, desktop or laptop—to get serious computing done. But your iPhone is now so powerful, so capable, and so well equipped with features and apps that you can use it as your main computer.

This chapter shows you the moves you'll need to make to turn your iPhone into your main computer. We'll start with pro tricks for entering text quickly and accurately using the onscreen keyboard, which hides plenty of secrets most people miss. Then, for top-speed text entry, we'll connect a Bluetooth keyboard to your iPhone. That'll put you in good shape for the next topic: how to create business documents—Word documents, Excel spreadsheets, PowerPoint presentations, and PDF files—on your iPhone wherever you are.

After that, I'll show you how to use your iPhone first as a portable drive for your computer and then as a file server for your home network. Both these moves let you not only take your vital files with you wherever you go but also access them from any computer you're using.

At the end of the chapter, you'll learn how to develop power-user e-mail skills with the Mail app and how to give presentations directly from your iPhone.

Let's get started.

Project 19: Learn Pro Tricks for Entering Text Quickly and Accurately

Fans of full-size computers may scorn the iPhone's onscreen keyboard, but you can enter text quickly and accurately with it—provided you know its secrets. On the iPhone 4S, you can also dictate text using Siri.

At first glance, your iPhone's onscreen keyboard could hardly be simpler to use:

- Tap the letter you want to type.
- Tap the Shift button to get an uppercase letter.
- Tap the .?123 button to display the keyboard containing numbers and common symbols.

- From the numbers and common symbols keyboard, tap the # + = button to display the keyboard containing brackets, braces, comic-book expletive characters (#%^&!), and so on.
- Tap the ABC button when you need the letter keys again.

But the onscreen keyboard also has a bunch of hidden tricks that can save you taps, time, and trouble. Read on...

Enter an Accented or Alternate Character

Tap and hold the base character until a pop-up panel appears, and then tap the character you want. For example, tap and hold the E until the panel shown here appears, and then tap the character needed.

Enter an En Dash (–) or an Em Dash (—)

Tap and hold the hyphen key, and then tap the en dash (–) or em dash (—) on the pop-up panel. An *en dash* is a dash the width of an N character in the font, and an *em dash* is a dash the width of an M character, which is substantially wider.

From this pop-up panel, you can also enter a bullet instead of having to go to the # + = screen.

Type a Period Quickly

To type a period quickly, tap the spacebar twice in quick succession.

If this doesn't work, you need to turn this feature on. See the section "Turn On All the Automatic Correction Features," coming right up.

Enter Other Domains Than .com

When you're using Safari, the onscreen keyboard has a .com button you can tap to enter the .com domain easily. To enter other widely used domains, tap and hold the .com button, and then tap the domain on the pop-up panel (shown here).

Enter the Domain for an E-Mail Address

To enter the domain for an e-mail address, tap and hold the . (period) key, and then tap the domain on the pop-up panel (shown here).

Enter Punctuation and Switch Straight Back to the Letter Keyboard

Often, you'll need to enter a single punctuation character, and then go back to typing letters. To do this, tap the .?123 button, but don't remove your finger from the screen. Slide your finger across to the punctuation key, and then take your finger off the screen. The iPhone enters the character and displays the letter keyboard again.

Keep Typing and Let Automatic Correction Fix Your Typos

Your iPhone's automatic correction features (discussed next) sort out many typos. So if you find you've made a typo halfway through a word, you're usually better off to keep typing and accept the automatic correction (as shown here) than to go back and fix the typo.

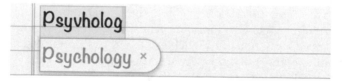

Turn On All the Automatic Correction Features

Your iPhone has a handful of automatic correction features to help you enter text faster and more accurately. To open the Keyboard screen and choose settings, follow these steps:

1. Press the Home button to display the Home screen.
2. Tap the Settings icon to display the Settings screen.
3. Scroll down to the third box, the one that starts with the General button.
4. Tap the General button to display the General screen.
5. Scroll down to the sixth box, the box that starts with the Date & Time button.

6. Tap the Keyboard button to display the Keyboard screen. The left screen in the next illustration shows the upper part of the Keyboard screen. The right screen shows the lower part of the Keyboard screen.

7. Set the Auto-Capitalization switch to the On position if you want your iPhone to automatically capitalize the first word of a new sentence or a new paragraph.
8. Set the Auto-Correction switch to the On position if you want to use automatic corrections and text shortcuts. These are usually helpful.
9. Set the Check Spelling switch to the On position if you want the iPhone to check your spelling and query apparently misspelled words.
10. Set the Enable Caps Lock switch to the On position if you want to be able to turn on Caps Lock by double-tapping the Shift button. This is usually helpful unless you find yourself turning on Caps Lock by accident.
11. Set the "." Shortcut switch to the On position if you want to be able to type a period by tapping the spacebar twice, as described earlier in this chapter. This shortcut is usually helpful.
12. Leave the Keyboard screen displayed so that you can set up text shortcuts, as described next.

Create Text Shortcuts

If you've used Microsoft Word or another word processor, you're familiar with the AutoCorrect feature, which automatically fixes typos and expands shortcuts you've defined (for example, expanding **myadd** to your full mailing address). Your iPhone has a similar feature, and you can speed up your typing by defining shortcuts like this:

1. Display the Keyboard screen as explained in the previous list.
2. At the bottom of the screen, tap the Add New Shortcut button to display the Shortcut screen (shown here).
3. Type the replacement word or phrase in the Phrase box.
4. Type the shortcut in the Shortcut box.
5. Tap the Save button.
6. Repeat steps 2–5 for each other shortcut you want to create.
7. When you've finished creating shortcuts and choosing keyboard settings, tap the Keyboard button to go back to the Keyboard screen.

	12:04 PM	
Keyboard	**Shortcut**	Save
Phrase	management	
Shortcut	mgt	

Create a shortcut that will automatically expand into the word or phrase as you type.

DOUBLE GEEKERY

Dictate Your Terms to Siri

I'm sure you know that Siri can be a great way to enter text on your iPhone quickly and accurately. But you can greatly boost your speed and accuracy by using the terms that Siri recognizes.

To dictate text to Siri, you simply position the insertion point in the note, e-mail message, or other document you're writing, summon the Siri microphone, and then say the words you want Siri to write down for you. When you stop speaking, Siri processes your input and then writes down the text that the servers in the Apple data center have understood.

To insert punctuation, just say the punctuation word in the stream of text. You don't need to warn Siri that you're about to use a punctuation term. These are the punctuation terms you can use:

- "Period" or "full stop"
- "Comma"
- "Semicolon"
- "Colon"
- "Exclamation point" or "exclamation mark"
- "Inverted exclamation mark"
- "Question mark"
- "Inverted question mark"

- "Hyphen"
- "Dash" or "en dash" (–)
- "Em dash" (—)
- "Underscore"
- "Open parenthesis" and "close parenthesis"
- "Open bracket" and "close bracket"
- "Ampersand"
- "Asterisk"

Here is a list of symbols you can tell Siri to enter:

- "At sign"
- "Copyright sign"
- "Registered sign"
- "Pound sign" or "hash mark" (#)
- "Dollar sign"
- "Cent sign"
- "Euro sign" (€)
- "Pound sterling sign" (£)
- "Yen sign" (¥)
- "Percent sign"
- "Greater-than sign" and "less-than sign"
- "Forward slash" and "backslash"
- "Vertical bar" (|)
- "Caret" (^)

To tell Siri how to format and lay out text, use these commands:

- **"New line"**   This gives a single line, with no blank line between paragraphs.
- **"New paragraph"**   This gives two lines, so you get a blank line between paragraphs.
- **"Cap"**   This makes Siri apply an initial capital to the following word. For example, "cap cheese" produces *Cheese*.
- **"No caps"**   This prevents Siri from applying an initial capital to a word that would normally get one. For example, "he's no caps Russian" produces *he's russian*.
- **"Caps on" and "Caps off"**   "Caps on" turns Caps Lock on, making everything you dictate all caps until you say "caps off."
- **"No caps on" and "No caps off"**   "No caps on" locks caps off, making everything you dictate all lowercase until you say "no caps off," after which normal capitalization resumes.
- **"Open quotes" and "Close quotes"**   For example, "open quotes cap hello exclamation point close quotes she said period" produces *"Hello!" she said*.
- **"Spacebar"**   This forces Siri to insert a space between words that it would otherwise put a hyphen between. For example, "this document is up spacebar to spacebar date" produces *this document is up to date* (instead of *up-to-date*).
- **"No space"**   This prevents Siri from inserting a space between words. For example, if you needed to enter the product name *BovineEmulator*, you could say "cap bovine no space cap emulator." You can also say "no space on" to turn on no-spacing until you say "no space off" to turn it off again.
- **"Dot"**   This puts a period between two words—for example, "Amazon dot com" produces *Amazon.com*.
- **"Point"**   This puts a period between numbers—for example, "two point five times as likely to succeed" gives *2.5 times as likely to succeed*.

Finally, you can also enter the most common emoticons by saying "smiley face," "frown face," and "wink face."

Project 20: Connect a Bluetooth Keyboard to Your iPhone

Your iPhone's onscreen keyboard is as good as Apple can make it, and Siri's dictation capabilities can be a great help in writing reminders, notes, e-mails, and text messages without needing to get your fingers dirty (or clean).

But when you need to enter serious amounts of text on your iPhone, there's no substitute for a hardware keyboard. By connecting a Bluetooth keyboard, you enable yourself to enter text at full speed in any app that requires it—Mail, Notes, the iWork apps, Documents To Go, or whatever.

 If you like Apple products, you've no doubt checked out the iPad—and maybe seen Apple's iPad Keyboard Dock accessory, which connects a hardware keyboard via the Dock Connector port. You might wonder if the iPad Keyboard Dock works with the iPhone too, as it would be handy to mount the iPhone on the keyboard as you work. But this accessory doesn't work with the iPhone. You need to use a Bluetooth keyboard instead.

Before you can connect your Bluetooth keyboard to your iPhone, you must do two things:

- **Turn on Bluetooth**   To save power, your iPhone keeps Bluetooth turned off until you need it.
- **Pair the keyboard with your iPhone**   Pairing is a one-time procedure that introduces the keyboard to your iPhone and sets them up to work together. Pairing helps ensure that only approved Bluetooth devices can connect to your iPhone.

Turn On Bluetooth

To open the Bluetooth screen in the Settings app and turn on Bluetooth, follow these steps:

1. Tap the General button on the Settings screen to display the General screen.
2. Tap the Bluetooth button to display the Bluetooth screen (shown on the left in Figure 3-1 with Bluetooth turned on).
3. Tap the Bluetooth switch and move it to the On position.

Pair Your Bluetooth Keyboard

To pair your Bluetooth keyboard, follow these steps:

1. Turn Bluetooth on, as described in the previous section.
2. Put the Bluetooth keyboard into Pairing mode. How you do this depends on the keyboard, but it usually involves a magic press on the power button—for example, pressing and holding the power button until red and blue lights start flashing.

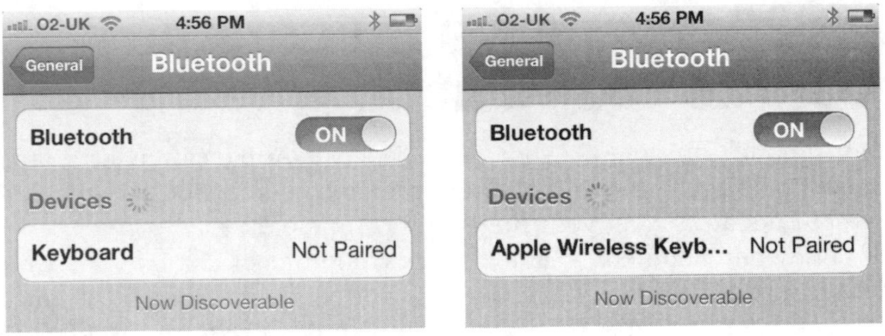

FIGURE 3-1 On the Bluetooth screen in the Settings app (left), tap the Bluetooth switch and move it to the On position. Your iPhone then detects nearby Bluetooth devices in pairing mode (right).

3. When the keyboard's button appears in the Devices list, showing Not Paired (see the right screen in Figure 3-1), tap the button to connect the keyboard.
4. Your iPhone then prompts you to type on the keyboard a pairing code, as shown in the left screen in Figure 3-2. Type the code and press ENTER or RETURN, and the keyboard and iPhone establish the pairing.

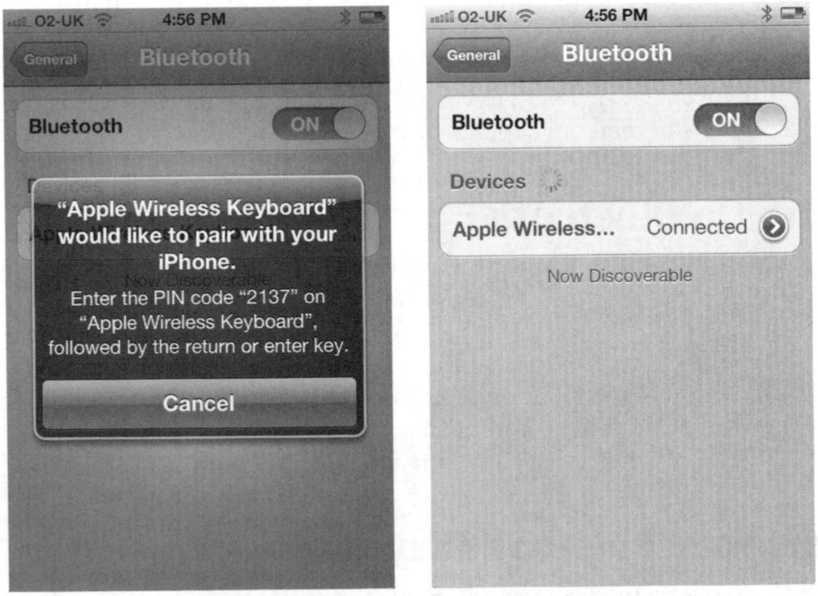

FIGURE 3-2 Your iPhone prompts you to enter a pairing code on your Bluetooth keyboard (left). After your iPhone establishes the pairing, it connects the Bluetooth keyboard (right).

After pairing the keyboard, your iPhone connects it automatically (see the right screen in Figure 3-2), on the assumption that you want to use the keyboard you're pairing. For subsequent use, connect the keyboard as described in the upcoming section "Connect Your Bluetooth Keyboard Again."

Disconnect Your Bluetooth Keyboard

To disconnect your Bluetooth keyboard from your iPhone, turn the keyboard off. Your iPhone then shows the keyboard as Not Connected in the Devices list on the Bluetooth screen.

You can also disconnect the keyboard by turning off Bluetooth on your iPhone.

 When you're not using Bluetooth, turn it off to save power and extend your iPhone's battery life.

Connect Your Bluetooth Keyboard Again

After you have paired your Bluetooth keyboard, you can quickly connect it to your iPhone again by moving it within Bluetooth range of your iPhone and turning it on. As long as your iPhone's Bluetooth is turned on, the iPhone connects to the keyboard, and you can start using it within a few seconds.

Make Your iPhone Forget Your Bluetooth Keyboard

When you no longer need to use your Bluetooth keyboard with your iPhone, you can tell your iPhone to forget it. Follow these steps:

1. On the Bluetooth screen, tap the keyboard's > button to display the control screen for the keyboard, shown here.
2. Tap the Forget This Device button. Your iPhone displays this confirmation dialog box.

3. Tap the Forget Device button. Your iPhone forgets the device, and then displays the Bluetooth screen again.

Change the Keyboard Layout
for a Bluetooth Keyboard

Connecting a Bluetooth keyboard to your iPhone is a great way to enter text quickly. And if you're used to a different layout on the keyboard, such as a European layout or the optimized Dvorak layout, you can switch the keyboard to use that layout.

If you need to use a different keyboard layout than the one you currently get on either the onscreen keyboard or an external keyboard you've connected, follow these steps to change the keyboard layout:

1. Press the Home button to display the Home screen.
2. Tap the Settings icon to display the Settings screen.
3. Scroll down to the third box, the one that starts with the General button.
4. Tap the General button to display the General screen.
5. Scroll down to the sixth box, the box that starts with the Date & Time button.
6. Tap the Keyboard button to display the Keyboard screen.
7. Tap the International Keyboards button to display the Keyboards screen, shown here with a single keyboard added.

8. Tap the top button. This button's name depends on which keyboard you're using—for example, English. The screen for the keyboard appears.
9. In the Software Keyboard Layout box, tap the layout you want for the onscreen keyboard—for example, QWERTY.
10. In the Hardware Keyboard Layout box, tap the button for the keyboard layout you want for your external keyboard—for example, Dvorak.
11. Tap the Keyboards button to return to the Keyboards screen.
12. Tap the Keyboard button to return to the Keyboard screen.

From the Keyboards screen, you can also add a new keyboard by tapping the Add New Keyboard button. But if you simply need to change the keyboard you're using with the iPhone, change the existing one rather than adding another keyboard.

Project 21: Create and Share Business Documents

If you're using your iPhone as your main computer, you'll likely need to create business documents on it. In this project, I'll show you how to create and edit essential types of business documents—everything from killer memos and coldly calculating spreadsheets through drop-dead presentations and professional PDF files.

Even if you create your business documents on your iPhone, you probably won't want to keep them there. So I'll walk you through the different means you can use to share documents between your iPhone and your PC or Mac. You can copy documents back and forth using iTunes' File Sharing feature, attach documents to e-mail messages, transfer documents by using third-party apps, or share documents using Apple's iWork.com site.

 Your iPhone's operating system has built-in viewers for major file types, such as PDF files, Word documents (in both the .docx and .doc formats), Excel workbooks (both .xlsx and .xls), PowerPoint presentations (both .pptx and .ppt), rich-text format (RTF), plain text, and HTML. Various apps can access these viewers—for example, if you receive a Word document attached to an e-mail message, Mail can open the document in a viewer so you can see its contents. But you have to buy third-party apps to edit these document types.

Create Business Documents on Your iPhone

In this section, we'll look quickly at the main apps for creating business documents on the iPhone: word-processing documents, spreadsheets, presentations, and PDF files.

Given that Microsoft Office not only dominates the Windows market for Office documents but also has a hefty chunk of the Mac market, it's most likely you'll need to create your business documents in the Word, Excel, and PowerPoint formats—so we'll start there. Next, we'll cover creating documents in the Pages, Numbers, and Keynote formats used by the apps in Apple's iWork suite. Finally, we'll look at how to create PDF files.

Create Documents in the Microsoft Office File Formats

To create documents in the Microsoft Office file formats on your iPhone, you have four main choices:

- **Documents To Go**   The basic version of Documents To Go can create Word documents and Excel spreadsheets and view PowerPoint presentations and iWork files. The advanced version, Documents To Go Premium, adds creating and editing PowerPoint presentations to the list. Figure 3-3 shows Documents To Go opening (left) and at work on a Word document (right).

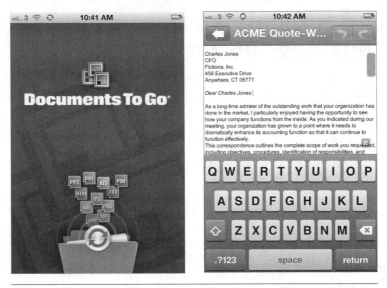

FIGURE 3-3 Documents To Go Premium can create and edit Word documents, Excel spreadsheets, and PowerPoint presentations.

- **Quickoffice** The basic version of Quickoffice can create Word documents and Excel spreadsheets. The pro version, Quickoffice Pro, can create and edit PowerPoint presentations as well. Figure 3-4 shows Quickoffice Pro creating a presentation (left) and a spreadsheet (right).

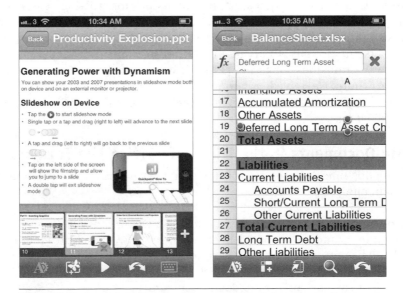

FIGURE 3-4 Quickoffice Pro can create Word documents, Excel spreadsheets, and PowerPoint presentations.

- **Google Docs** If you have an account with Google Docs (http://docs .google.com), you can log into it using Safari or another web browser on your iPhone, and then create word-processing documents, spreadsheets, and presentations in it. The interface is clumsy because of the iPhone's screen being small, but if you have a sure touch, it's workable. Figure 3-5 shows the iPhone creating a presentation in Google Docs.

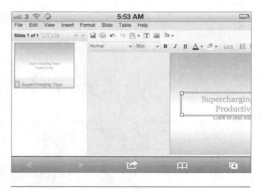

FIGURE 3-5 You can use Safari or another web browser to create and edit word-processing documents, spreadsheets, or presentations in Google Docs.

- **iWork** Pages, Numbers, and Keynote (discussed in the next section) can export files in the corresponding Microsoft Office formats. For example, from Numbers, you can export a spreadsheet in the Microsoft Excel format. See the nearby sidebar "Convert Your iWork Files to the Microsoft Office Formats" for details.

Documents To Go and Quickoffice are impressive apps, but they enable you to use only the most widely used formatting and objects (such as tables and shapes) when creating documents, spreadsheets, and presentations. Because of these limitations, and because the iPhone's screen offers only a small area to work in, you will normally do best to finish your documents on a computer rather than on the iPhone.

Create Documents in the iWork File Formats

If you need to create documents in the iWork file formats on your iPhone, look no farther than Apple's iWork apps. These apps are the iPhone versions of the full-scale Mac OS X applications:

- **Pages** Pages is an app for creating word-processing and layout documents. The left screen in Figure 3-6 shows Pages working on a document.
- **Numbers** Numbers is an app for creating spreadsheets. The right screen in Figure 3-6 shows a spreadsheet open in Numbers.
- **Keynote** Keynote is an app for creating and editing presentations. Figure 3-7 shows Keynote.

FIGURE 3-6 Use the Pages app (left) to create and edit word-processing documents on your iPhone or use the Numbers app (right) to create and edit spreadsheets.

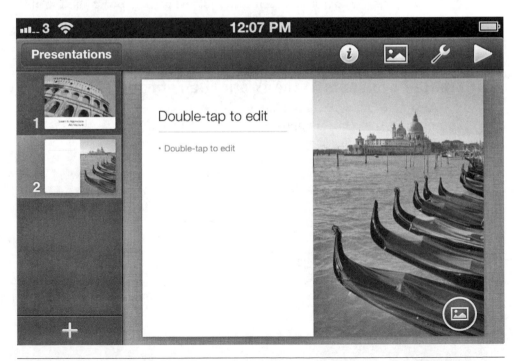

FIGURE 3-7 Use the Keynote app to create and edit presentations on your iPhone. You can export presentations in Microsoft PowerPoint format if necessary.

Convert Your iWork Files to the Microsoft Office Formats

The three iWork apps—Pages, Numbers, and Keynote—are great for working on the iPhone (and even better on the iPad, because it gives you so much more screen space). But if you or your colleagues use Microsoft Office on your computers, you'll need to convert the iWork files you create to their Office equivalents. To convert the files, you use the Share And Print feature in the iWork apps.

To convert a file using the Share And Print feature, follow these steps on your iPhone:

1. Open the app to which the document belongs. I'll use Pages for this example.
2. If the app launches with a document open other than the document you want to convert, tap the Documents button, the Spreadsheets button, or the Presentations button in the upper-left corner of the screen to go back to the Document Manager screen. This is the screen that shows the contents of the Documents folder, the Spreadsheets folder, or the Presentations folder.
3. Tap the document you want to convert. The app opens the document.
4. Tap the Tools button (the button with the wrench icon in the upper-right corner of the screen) to display the Tools screen (shown on the left in the next illustration).

Tools	Done		Share and Print	Done
🔗 Share and Print	>		✉ Email Document	
🔍 Find			🖨 Print	>
📄 Document Setup			Share via iWork.com	
📏 Ruler			♫ Send to iTunes	
⚙ Settings	>		☁ Copy to iDisk	
❓ Help	>		👥 Copy to WebDAV	

5. Tap the Share And Print button to display the Share And Print screen (shown on the right in the illustration).

6. Tap the Email Document button or the Send To iTunes button, as appropriate. Your iPhone displays a screen for choosing the format of the document. The left screen in the next illustration shows the Email Document screen, which Pages displays when you choose to e-mail a document. The Email Spreadsheet screen in Numbers, the Email Presentation screen in Keynote, and the Choose Format screen (for sending to iTunes) offer similar choices.

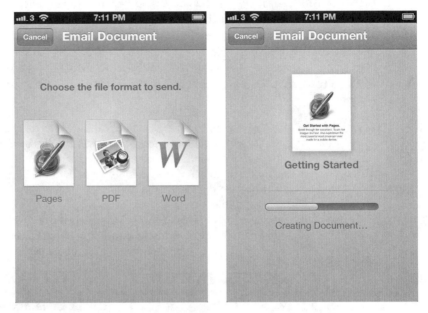

7. Tap the format to use for the exported file:
   - **Native format** Tap the Pages button, the Numbers button, or the Keynote button to keep the document in its native format.
   - **PDF** Tap the PDF button to create a Portable Document Format file for viewing on any computer (but not for editing).
   - **Office format** Tap the Word button (from Pages), the Excel button (from Numbers), or the PowerPoint button (from Numbers).
8. The app exports the file in the format you chose, and then displays the document again.

Create PDF Files

Creating documents, spreadsheets, or presentations is helpful, but sometimes you may need to create PDF files on your iPhone so that you can give your clients fully laid out documents that they can't change.

 If you're using Pages, Numbers, or Keynote, you can create a PDF file by exporting the document. See the preceding sidebar "Convert Your iWork Files to the Microsoft Office Formats" for details.

When you need to create PDF files, try these two apps:

- **Adobe CreatePDF**   CreatePDF from Adobe, the company behind the PDF file format, enables you to take a document from a file-storage area and turn it into a PDF file. CreatePDF is a little clumsy, because it doesn't have a file browser for picking the document from which to create the PDF file: Instead, you have to start from the file-storage area of the app the document is in, and then use the Open In command to open it in CreatePDF, as shown on the left in Figure 3-8. But once you've picked the document, the conversion to PDF (shown on the right in Figure 3-8) runs smoothly.
- **Save2PDF**   Save2PDF is an app for creating and manipulating PDF files. Save2PDF's features include merging two or more PDF files into a single file and adding extra pages to an existing document. For example, if you have a PDF file that contains a standard contract, you can add to it an extra page that turns it into a customized version.

Share Documents with Your PC or Mac

In this section, we'll look at how to share documents between your iPhone and your computer. We'll cover iTunes' File Sharing feature, look into transferring documents via e-mail, and discuss three third-party apps that can transfer documents.

FIGURE 3-8   To create a PDF file using CreatePDF, you use the Open In command (left). CreatePDF then converts the file to a PDF.

Share Documents Using iTunes' File Sharing

If you use iTunes rather than iCloud to sync your iPhone, you can use iTunes'
File Sharing feature to put documents on the iPhone from your computer or copy
documents from your iPhone to your computer. This is the most direct way of shifting
files from point A to point B.

To transfer documents by using File Sharing, follow these steps:

1. Connect your iPhone to the computer as usual.
2. If the computer doesn't automatically launch or activate iTunes, launch or
   activate iTunes yourself.
3. In the Source list, click your iPhone's entry to display its control screens.
4. Click the Apps tab to display your iPhone's apps and files.
5. Scroll down to the File Sharing area (see Figure 3-9).
6. In the Apps list, click the app to which you want to transfer the files. The list of
   files for that app appears in the Documents pane to the right.
7. To add documents to the app, follow these steps:
   a. Click the Add button to display the Open dialog box.
   b. Navigate to and select the document or documents you want to add.
   c. Click the OK button (on Windows) or the Open button (on the Mac).

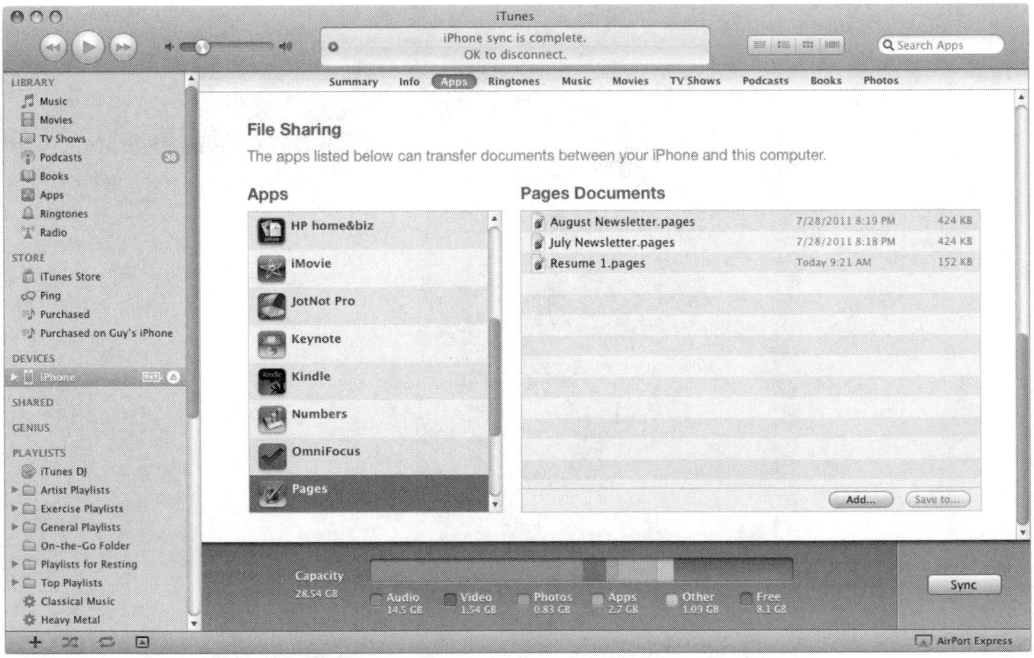

FIGURE 3-9 The File Sharing area on the Apps tab in the iTunes control screens for an iPhone
lists the apps that can transfer files. Click an app to see its files.

8. To copy documents from the app to the computer, follow these steps:
   a. Click the Save To button to display the iTunes dialog box (on Windows) or the Choose A File: iTunes dialog box (on the Mac).
   b. Navigate to the folder in which you want to save the document.
   c. Click the Select Folder button (on Windows) or the Open button (on the Mac).

File transfer generally run pretty quickly, as USB 2.0 can handle up to 480 megabits per second (Mbps)—but if you're transferring many large files, it'll take a while. (And if you're using a USB 1.*x* port, its 12 Mbps limit will make things much slower.)

Transfer Documents via E-Mail

When you need to get documents onto your iPhone quickly, you can simply e-mail them to an account on it. You can then open a document directly from the e-mail message into either one of the iPhone's viewers or into whichever app you want to use to work on the document.

E-mail may seem like a clumsy solution to document transfer, but it's quick and effective unless the document is too big to go through e-mail servers. E-mail is especially useful when the document is on somebody else's computer rather than the computer with which you normally sync your iPhone.

And you don't need me to point out that you can use Mail to send a document back after you've edited it, or send it along to the next person who needs to deal with it.

Copy a Document from a Message to an App's Storage Area   To get a document out of an e-mail message and into an app's storage area, follow these steps:

1. In the message list, tap the message to display its contents.
2. Tap and hold the document's button in the message until Mail displays a menu (shown on the left in Figure 3-10).
3. If you want to open the basic viewer for the document, tap the Quick Look button; normally, though, you'll do better to open the document in an app. If you want to open the document in the default app (in this example, Pages), tap the Open In "*App*" button (where *App* is the app's name). Otherwise, tap the Open In button to display the Open In menu (shown on the right in Figure 3-10), and then tap the app you want to use.

That's the most efficient way to copy the document from the message and get it into the app. But what you'll probably want to do often is view the contents of the document so that you can decide which app to open it in. For example, if you receive a Word document on your iPhone, you may want to bring it into Pages so that you can use Pages' streamlined layout tools. But if you simply want to edit the document as a Word document, you'll do better to open the document in Documents To Go or a similar app that can maintain the Word document format.

FIGURE 3-10 Tap and hold a document's button in an e-mail message until the menu for opening the document appears (left). To use a different app, tap the Open In button to display the Open In menu (right), and then tap the app you want.

View a Document and Decide Which App to Open It In   To view a document and then decide which app to open it in, follow these steps:

1. In the message list, tap the message to display its contents.
2. Tap the button for the attached document you want to open. Your iPhone displays the document in the viewer.
3. Tap the action button (the button with a curving arrow in the upper-right corner of the screen) to display the menu for opening the document in the default app, opening it in another app, or printing it. The left screen in Figure 3-11 shows a PDF document, for which iBooks is the default app.
4. If you want to use the default app, tap its button to open the document in it. Otherwise, tap the Open In button to display the list of apps that can open the document. The right screen in Figure 3-11 shows an example of this list.
5. Tap the app in which you want to open the document.

This approach leaves the document open in the viewer in Mail. So when you go back to Mail, tap the Message button to close the viewer and return to the message.

Delete the Document from Mail if Necessary   Once you've opened an attached document in another app, that app stores a copy of the document in its storage area.

FIGURE 3-11   To copy a document from an e-mail message, open it in the viewer. From the action menu (left), you can open the document in its default app. To use another app, tap the Open In button, and then tap the app to use (right).

You can now delete the e-mail message and the attached document if necessary; the copy of the document that you've added to the other app's storage area remains unaffected.

 If you attach a picture to an e-mail message, the recipient can save the picture to his or her iPhone's Photos storage area. But if you attach a music file or video file, the recipient can only play it in the viewer or add it to third-party apps that handle media file types, not add it to the iPhone's Music storage area.

Transfer Documents Using Third-Party Apps

If you need more direct or wider-ranging access to the iPhone's file system than iTunes provides, you'll need to use a third-party app instead. This section introduces you to three of the most widely useful apps at this writing: Air Sharing, FileApp Pro (with or without DiskAid), and Documents To Go.

Transfer Documents Using Air Sharing   Air Sharing is an app for transferring documents to and from the iPhone and viewing them on the device. Air Sharing enables you to connect your computer to your iPhone via a wireless network connection and comes in three different versions:

- **Air Sharing**   Air Sharing is the basic version of the app for the iPhone. You can mount the iPhone as a drive on a PC or Mac, transfer files both ways, and view or e-mail documents in the formats the iOS viewer supports. Figure 3-12 shows a Finder window displaying the contents of an iPhone mounted as a drive using Air Sharing.
- **Air Sharing Pro**   Air Sharing Pro adds abilities such as connecting to a Windows PC running a companion program, mounting remote file systems, opening and creating Zip files, and downloading files from the Web.
- **Air Sharing HD**   Air Sharing HD is the iPad version of Air Sharing Pro and provides similar features at the larger screen size.

 For instructions on connecting a PC or Mac to your iPhone via Air Sharing, see Project 23, "Use Your iPhone as a File Server for Your Household," later in this chapter.

Transfer Documents Using FileApp Pro   Like Air Sharing, FileApp Pro is an app for transferring documents to and from your iPhone and for viewing documents on the devices. With FileApp Pro, you can connect to the iPhone either via the USB cable

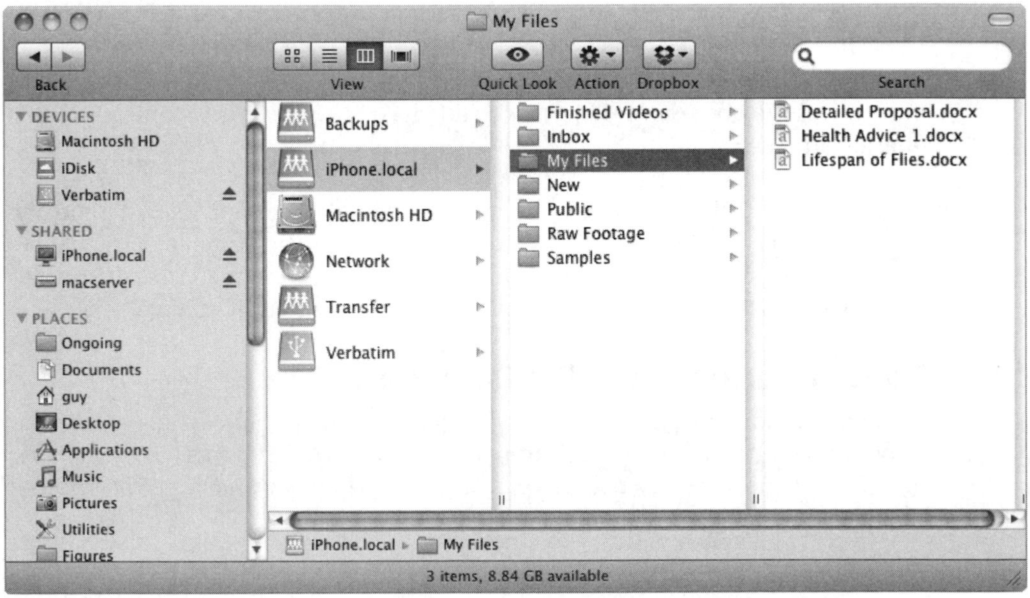

FIGURE 3-12   With Air Sharing, you can mount an iPhone as a drive on your computer so that you can easily transfer files.

(which is good for speed) or via a wireless network connection (which is good for flexibility). To connect via USB, you need to either use iTunes' File Sharing feature or run the DiskAid program from DigiDNA ($24.90; www.digidna.net) on your PC or Mac.

The left image in Figure 3-13 shows the FileApp Pro interface for manipulating folders on the iPhone. The right image in Figure 3-13 shows the Sharing screen of FileApp Pro. You tap the USB button or the WIFI button to choose the means of sharing, and then tap the button for the operating system you're using—Win 7, Win Vista, Win XP, or Mac OS X—to display instructions for connecting.

You can transfer documents by using the FileApp Pro entry in the File Sharing area of the Apps tab in iTunes, but if you want to transfer many files easily and choose the folders to put them in, it's worth getting DiskAid and installing it on your computer. Once you've set up sharing on the iPhone, you can connect via DiskAid and transfer files easily back and forth. Figure 3-14 shows DiskAid in action.

 DiskAid is a handy tool if you want to simply store files on your iPhone—for example, to transfer them from one computer to another—rather than open the files on the iPhone. With DiskAid, you can create your own folders on the iPhone, enabling you to use it as an external disk.

Transfer Documents Using Documents To Go   If you need to work extensively with Microsoft Office documents—for example, Word documents or Excel workbooks—you'll

FIGURE 3-13   FileApp Pro lets you create and manipulate folders easily on your iPhone (left) and choose between USB and Wi-Fi connections (right).

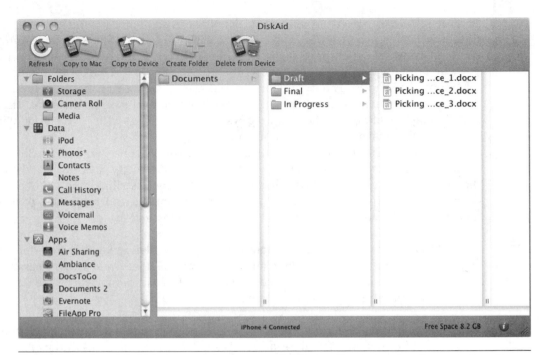

FIGURE 3-14   DiskAid is a companion program for FileApp Pro that makes it easy to transfer files to and from your iPhone. You can also use DiskAid on its own.

probably find the iWork apps too cumbersome. Instead of struggling with frustrating conversions, get a third-party program that can handle the main Office file formats without having to translate them.

As discussed earlier in this chapter, the main choices for creating and editing Microsoft Office documents directly on the iPhone are Documents To Go and Quickoffice. At this writing, Documents To Go seems the stronger of the two, especially as it has good features for transferring documents between your computer and your iPhone.

You can load documents into Documents To Go by using the Documents To Go entry in the File Sharing area of the Apps tab in iTunes, but for regular use, download the free companion desktop program that runs on your PC or Mac to synchronize documents with the iPhone. To get the program, go to the Documents To Go for iPhone page on the DataViz web site (www.dataviz.com/DTG_iphone.html), and then click the Download Win button or the Download Mac button. Once you've installed the program, you go through a HotSync setup process to pair your iPhone with the desktop program. You can then use the desktop program to transfer files to and from your iPhone (see Figure 3-15).

 Documents To Go Premium can access documents in an online storage account such as Google Docs, Box.net, Dropbox, or iDisk.

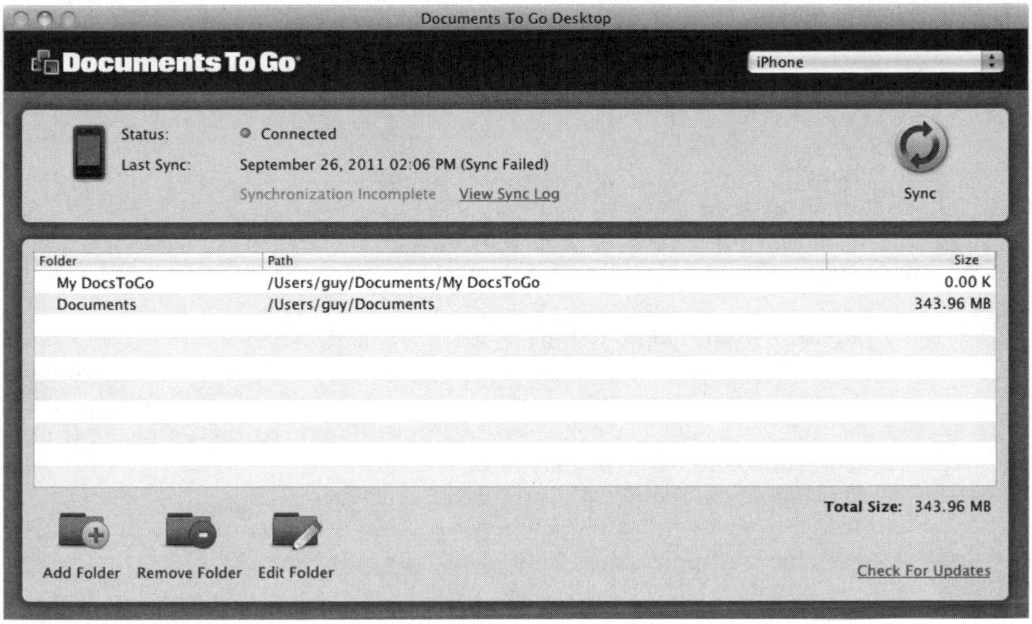

FIGURE 3-15   The Documents To Go desktop program runs on your computer and connects to your iPhone.

Project 22: Use Your iPhone as a Portable Drive

If you have files you need to keep with you at all times, you can store them on your iPhone and use it as a portable drive. This is great both for carrying files you need to be able to access at any moment or from any computer and for making backups of your most critical files.

You can also use your iPhone as a handy way to transfer large files from one computer to another.

In this section, I'll show you how to copy files to and from your iPhone. We'll start with iTunes' File Sharing feature, which you use to copy files to and from the iPhone's storage area for a particular app. We'll then move on to third-party programs that let you store files wherever you want in your iPhone's file system.

Storing files on your iPhone is a great way to keep them available to you at all times, but make sure you back them up to your computer or an online storage account as well. Otherwise, if your iPhone gets lost, stolen, or merely smashed, you'll lose forever any files stored only on the iPhone.

Copy Files to Your iPhone Using iTunes' File Sharing

The first way to copy files to and from your iPhone is by using iTunes' File Sharing feature. File Sharing enables you to put a file into the file-storage area for a particular app rather than into a folder of your own choice. See the nearby sidebar "Understand the iPhone's Separate Storage Areas for Documents" for an explanation of how these file-storage areas work.

To transfer documents to and from your iPhone by using File Sharing, follow these steps:

1. Connect your iPhone to the computer as usual.
2. If the computer doesn't automatically launch or activate iTunes, launch or activate iTunes yourself.
3. In the Source list, click the entry for your iPhone to display its control screens.
4. Click the Apps tab to display its contents.
5. Scroll down to the File Sharing area (see Figure 3-16).
6. In the Apps list, click the app to which you want to transfer the files. The list of files for that app appears in the Documents pane to the right.

DOUBLE GEEKERY

Understand the iPhone's Separate Storage Areas for Documents

For security, the iPhone's file system, iOS, gives each app a separate storage area for documents. iOS largely confines each app to its own storage area and prevents it from accessing the storage areas of other apps. This security measure both protects against malware and prevents one app from making unwanted changes to another app's data files.

For example, if you have the Pages app on your iPhone, you can use File Sharing to transfer a Pages document from your Mac to your iPhone. Once the Pages document is on your iPhone, you can launch Pages and then open the document. But you can't open the document in another app, because it's stored in Pages' storage area.

The exception is apps that can receive incoming files, such as Mail and Safari. These apps can provide those files to other apps. For example, if you receive a Word document attached to an e-mail message on your iPhone, you can choose to open that document in Pages or another app that can handle Word documents. Mail makes the document available to Pages or the app you choose.

When you open the document in Pages, your iPhone copies the document to Pages' storage area. You can then use Pages to open that new copy from Pages' storage area. The original attached document remains attached to the message in Mail, and you can copy it to another app's storage area if you need to.

FIGURE 3-16 The File Sharing area on the Apps tab in the iTunes control screens for an iPhone lists the apps that can transfer files. Click an app to see its files.

7. To add documents to the app, follow these steps:
   a. Click the Add button to display the iTunes dialog box (on Windows) or the Choose A File: iTunes dialog box (on the Mac).
   b. Navigate to and select the document or documents you want to add.
   c. Click the OK button (on Windows) or the Choose button (on the Mac).
8. To copy documents from the app to the computer, follow these steps:
   a. Click the Save To button to display the iTunes dialog box (on Windows) or the Choose A File: iTunes dialog box (on the Mac).
   b. Navigate to the folder in which you want to save the document.
   c. Click the Select Folder button (on Windows) or the Choose button (on the Mac).
9. Click the Sync button to run the synchronization.

File transfers generally run pretty quickly, as USB 2.0 can handle up to 480 Mbps—but if you're transferring many large files, it'll take a while.

Find a Suitable Program for Transferring Other Files to and from Your iPhone

When you need full-on access to your iPhone's file system, iTunes' File Sharing isn't enough. Instead, you need to use a third-party program that enables you to use your iPhone as an external drive.

This section shows you three such programs: DiskAid, Air Sharing, and PhoneView. You can find others on the Web or in the App Store (which you can access via iTunes on your computer or via the App Store application on your iPhone).

DiskAid (Windows and Mac OS X)

DiskAid from DigiDNA ($24.90; www.digidna.net/diskaid) is a utility that lets you mount your iPhone as an external disk. Figure 3-14 (earlier in this chapter) shows DiskAid at work on a Mac.

DiskAid's toolbar buttons let you easily create folders, copy items to and from the device, and delete items from the device. But you can also simply drag files and folders from a Windows Explorer window or a Finder window to the DiskAid window to add them to the device.

 DiskAid includes a feature called TuneAid, which you can use to recover your songs from your iPhone to your computer—for example, after your computer's hard drive gives up the ghost and you have to replace it.

Air Sharing (Windows and Mac OS X)

Air Sharing from Avatron Software (www.avatron.com), which you can buy from the App Store for $6.99, lets you access your iPhone across a wireless network connection rather than the USB connection that most other programs require. Not having to connect the device to your computer is an advantage, but you get slower file transfers than via USB, and your iPhone doesn't get to recharge while you're using Air Sharing unless you plug it into the Apple USB Power Adapter.

If Air Sharing suits you, you may want to upgrade to the Universal version ($9.99), which includes a wider variety of file operations (such as creating new folders and zipping and unzipping files), the ability to mount remote file servers (such as iCloud and Dropbox), and the ability to print to certain printers.

 For instructions on setting up Air Sharing on your iPhone and connecting to it from your PC or Mac, see Project 23, "Use Your iPhone as a File Server for Your Household," later in this chapter.

PhoneView (Mac OS X Only)

PhoneView (see Figure 3-17) from Ecamm Network ($19.95; www.ecamm.com/mac/phoneview) lets you access your iPhone from your Mac. Ecamm provides a mostly

FIGURE 3-17   PhoneView lets you quickly access the contents of your iPhone to copy, add, or delete files.

functional trial edition, which gives you seven days to find out how well PhoneView suits your needs.

When you finish using PhoneView, quit it (for example, press ⌘-Q or choose PhoneView | Quit PhoneView). PhoneView closes its window and releases its grip on your iPhone's file system.

Project 23: Use Your iPhone as a File Server for Your Household

If your household has several computers, you probably need to share files among them. You can do so by sharing folders on one computer or another, but this works well only as long as each computer that's sharing files is turned on and running properly. Many people find that they're better off with a single computer or device sharing files, acting as a file server.

Using your iPhone as a file server means that you can keep all your important files on your iPhone, and take them with you wherever you go. Use your iPhone as a file server only for a small group of computers. The iPhone's wireless network connection transfers modest amounts of data plenty fast enough, but if you try to connect many computers to the iPhone at once, you'll get poor performance.

You can pay hundreds of dollars for a computer to use as a file server, or you can buy a network attached storage (NAS) device—in effect, a modest computer configured as a server on a network. But if you don't want to spend the money, you can turn your iPhone into a file server instead. All you need to do is install the right app, set it up to share files, and then connect your computers to it.

Keep your iPhone plugged in to a power source while you're using it as a server. And be sure to back up all the files you care about to avoid losing your data if you lose your iPhone.

You can get various apps that let you use your iPhone as a server, but at this writing the best bet is Air Sharing, which you've already met earlier in this chapter. In this section, I'll show you first how to set up Air Sharing on your iPhone, and then how to connect to your shared folders using your PC or Mac.

If you need to use several iPhones or iPod touches together as your file server, look at ServersMan (http://serversman.com/en/). This free app can be a great way of squeezing more use out of your old iPhones or iPod touches if you're too attached to them to sell them on eBay.

Set Up Air Sharing on Your iPhone

As discussed earlier in this chapter, Air Sharing comes in three different versions: Air Sharing, Air Sharing Pro, and Air Sharing HD. Air Sharing HD is the iPad version of Air Sharing, so for the iPhone you'll want either plain Air Sharing or Air Sharing Pro. I suggest you start with plain Air Sharing and upgrade to Air Sharing Pro only if you find you need the extra features it offers, such as connecting your iPhone to a Windows PC or mounting remote file systems on your iPhone.

After downloading Air Sharing and installing it on your iPhone, either by using the App Store on your iPhone or by synchronizing with iTunes, set up Air Sharing so that your computer can connect to it. Follow these steps:

1. On your iPhone, launch Air Sharing by tapping its icon on the Home screen. The My Documents screen appears, as shown on the left in Figure 3-18.
2. Tap the wrench icon in the lower-right corner to display the Settings screen (shown on the right in Figure 3-18).
3. Tap the Sharing button to display the Sharing screen (shown on the left in Figure 3-19).
4. Tap the Enabled switch and move it to the On position.

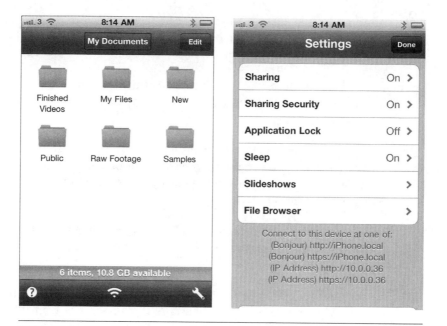

FIGURE 3-18   Tap the wrench icon in the lower-right corner of the My Documents screen (left) to display the Settings screen (right).

FIGURE 3-19   Turn on sharing by setting the Enabled switch on the Sharing screen (left) to the On position. On the Sharing Security screen (right), move the Require Password switch to the On position, and then enter the username and password for connecting.

After moving the Enabled switch on the Sharing screen to the On position, you can change the HTTP Port setting or the HTTPS Port setting if you need to. Normally, it's easiest to keep the default settings—port 80 for HTTP and port 443 for HTTPS.

5. Tap the Settings button to return to the Settings screen.
6. Tap the Sharing Security button to display the Sharing Security screen (shown on the right in Figure 3-19).

This section shows you how to implement a reasonable level of security on your iPhone for sharing. Air Sharing can provide not only password-free access but also public access to your iPhone, but normally you're better off providing only secured access to what you're sharing.

7. Tap the Require Password switch and move it to the On position.
8. Type the username and password to use for the connection. Each user will use the same username and password.
9. Tap the Public Access switch and move it to the Off position.
10. Tap the Settings button to return to the Settings screen. Look at the readout at the bottom giving the Bonjour addresses and IP addresses of your iPhone, and note the address you need—the non-https IP address for Windows, and the non-https Bonjour address for Mac OS X.
11. Tap the Done button to return to the My Documents screen.

Now that you've set up Air Sharing on your iPhone, you can connect to Air Sharing from your PC or Mac, as discussed next.

Connect to Air Sharing on Your iPhone from a PC

To connect to Air Sharing on your iPhone from a PC, follow these steps:

1. Choose Start | Computer to open a Computer window.
2. Click the Map Network Drive button on the toolbar to display the Map Network Drive dialog box (see Figure 3-20).
3. In the Drive drop-down list, choose the drive letter you want to map to your iPhone.
4. In the Folder text box, type **http://** and the IP address shown for your iPhone—for example, **http://10.0.0.36**.
5. Select the Reconnect At Logon check box if you want Windows to automatically reconnect the drive each time you log on. Unless you plan to run Air Sharing on your iPhone all the time, you're usually better off clearing this check box and establishing the connection manually when you need it.

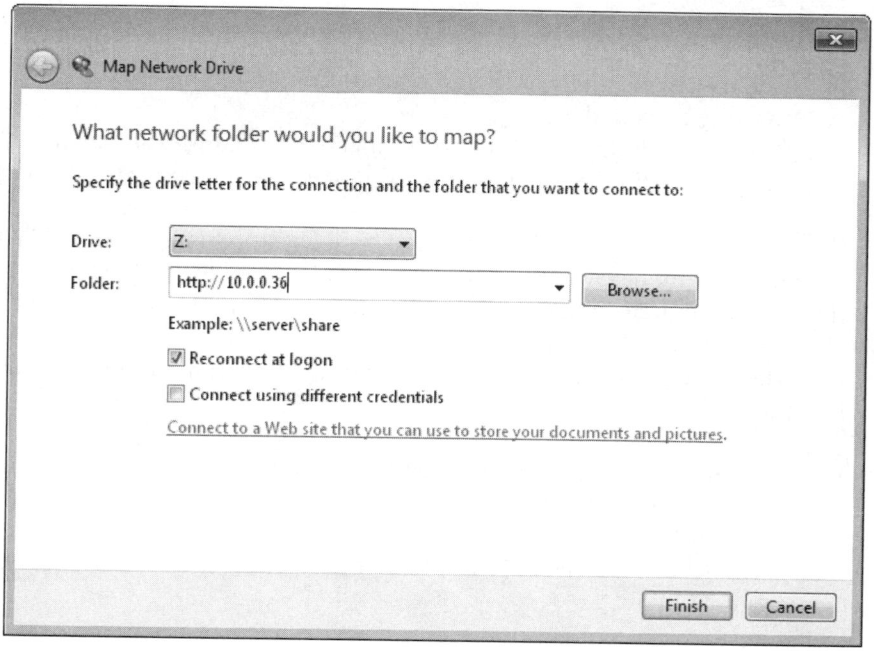

FIGURE 3-20   In the Map Network Drive dialog box, choose the drive letter to use, and then enter your iPhone's address in the Folder field.

6. Click the Finish button. Windows attempts to connect to your iPhone.
7. If you have set a username and password on Air Sharing, Windows prompts you to enter them, as shown here.

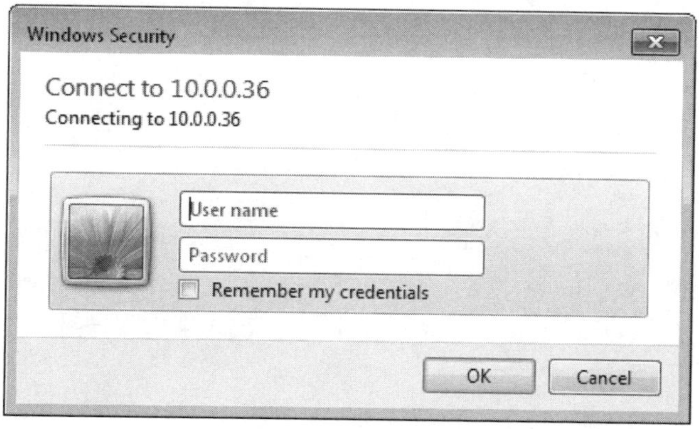

8. Type your username and password.
9. Select the Remember My Credentials check box if you want Windows to store the username and password for future use. If you're using your own PC, this is usually a good idea.
10. Click the OK button. Windows establishes the connection to your iPhone and displays a Windows Explorer window showing its contents (see Figure 3-21).

You can now work with your iPhone's file system using standard Windows Explorer techniques. For example, to create a new folder, right-click in open space in the document area, choose New | Folder from the context menu, type the name to give the folder, and then press ENTER.

When you finish using your iPhone from your PC, disconnect the network drive like this:

1. In the Windows Explorer window, click Computer in the address box to display the Computer window. Alternatively, choose Start | Computer to open a Computer window.
2. Right-click the drive representing your iPhone, and then click Disconnect on the context menu.

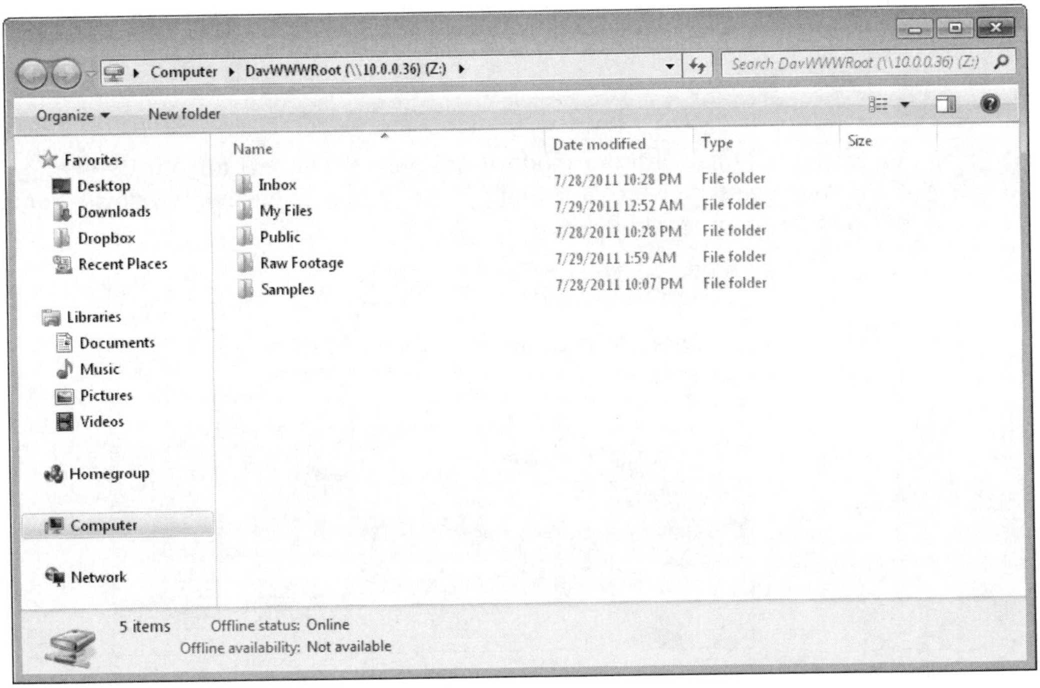

FIGURE 3-21 Windows opens a Windows Explorer window showing your iPhone's file system.

Connect to Air Sharing from Windows XP

If your PC is running Windows XP, you must have Service Pack 3 installed in order to connect to Air Sharing. If you're not sure which Service Pack your PC is running, click the Start button, right-click the My Computer icon, click Properties on the context menu, and then look at the System readout on the General tab of the System Properties dialog box.

There's also another complication: XP can't connect to the iPhone's shared directory. Instead, you must connect to a subdirectory—preferably the one you want to work in. If you use your iPhone mainly with a Windows XP PC, you'll probably want to set up your iPhone's file system with a subfolder that contains all your other folders.

Provided your PC has Service Pack 3 installed, connect like this:

1. Choose Start | My Computer to open a My Computer window.
2. Choose Tools | Map Network Drive to display the Map Network Drive dialog box.
3. In the Drive drop-down list, choose the drive letter you want to map.
4. In the Folder text box, type **http://**, the iPhone's IP address, a forward slash, and the name of a folder—for example, **http://10.0.0.36/Files**.
5. Click the Finish button.
6. If Windows XP displays a dialog box prompting you for your username and password, enter them, and click the OK button.

Connect to Air Sharing on Your iPhone from a Mac

To connect to Air Sharing on your iPhone from a Mac, follow these steps:

1. Click the desktop to activate the Finder.
2. Choose Go | Connect To Server or press ⌘-K to display the Connect To Server dialog box (shown here).

3. In the Server Address text box, type **http://** and the Bonjour address shown for your iPhone—for example, **http://iPhone.local**.

 Instead of typing the Bonjour address, you can type the IP address shown for your iPhone. But given that your Mac is running Bonjour anyway, the Bonjour address is usually a better choice. This is because your iPhone's Bonjour address remains the same unless you change your iPhone's name, whereas if your iPhone gets its IP address from a DHCP server (as is the normal setup), your iPhone will typically get a different IP address each time it connects to the DHCP server.

4. Click the Add (+) button if you want to add your iPhone to your list of servers. This is a good idea if you plan to access your iPhone frequently using this Mac.
5. Click the Connect button. The Finder attempts to connect to your iPhone.
6. If you have set a username and password on Air Sharing, Mac OS X prompts you to enter them, as shown here.

Enter your name and password for the server "iPhone.local".

Connect as: ○ Guest
             ● Registered User

Name: apple

Password: ••••••••

☐ Remember this password in my keychain

Cancel   Connect

7. Make sure the Registered User option button is selected.
8. Type your username and password.
9. Select the Remember This Password In My Keychain check box if you want your Mac to store the password for future use. When you're using your own Mac (as opposed to someone else's Mac), this is usually a good idea.
10. Click the Connect button. The Finder establishes the connection to your iPhone and displays your iPhone's contents in a Finder window.

You can now work with your iPhone's file system using the same techniques as for any other drive. For example, CTRL-click or right-click and then click New Folder on the context menu to create a new folder, as shown in Figure 3-22.

When you finish using your iPhone from the Mac, click the Disconnect button in the Finder window to disconnect the drive.

FIGURE 3-22   After connecting to your iPhone using Air Sharing, you can work with its file system using normal Finder techniques.

Project 24: Develop Power-User E-Mail Skills with Mail

Whether you use your iPhone for work, play, or both, you'll almost certainly want to work with e-mail on it. You'll be able to start using the Mail app easily enough, but you'll also find that it has plenty of power and lots of hidden secrets.

This section shows you ten ways to work faster and smarter with e-mail on your iPhone—everything from saving time by batch-editing your messages to working with drafts and changing the quote level in a message. We'll start by choosing five essential settings for Mail.

Choose Five Essential Settings

To make the Mail app behave your way, you can choose settings on the Mail, Contacts, Calendars screen. Start by opening the Mail, Contacts, Calendars screen like this:

1. Press the Home button to display the Home screen.
2. Tap the Settings icon to display the Settings screen.
3. Scroll down to the third box, the one that starts with the General button.
4. Tap the Mail, Contacts, Calendars button to display the Mail, Contacts, Calendars screen.

The left screen in Figure 3-23 shows the upper part of the Mail, Contacts, Calendars screen, and the right screen shows the lower part. At the top is the Accounts box, which contains a list of the accounts you've set up, and below that the Fetch New Data button and two boxes containing the Mail settings we'll work with in this section.

Many of the settings are straightforward, but this section shows you how to choose the five most important ones:

- Choose how many messages to show and how to preview them
- Set your default account for sending messages.
- Get your messages pushed to your iPhone
- Protect yourself against spam images
- Set up the signature you need

Choose How Many Messages to Show and How to Preview Them

At the top of the Mail box on the Mail, Contacts, Calendars screen (see the left screen in Figure 3-23), choose how many messages Mail should show and what kind of previews it should display for them:

- **Show**   Tap this button to display the Show screen, and then tap the button for the number of messages you want to see: 50 Recent Messages, 100 Recent Messages, 200 Recent Messages, 500 Recent Messages, or 1,000 Recent Messages. Unless you get a stack of e-mail, 50 Recent Messages is usually the best way to start. Tap the Mail, Contacts, Calendars button when you've made your choice.

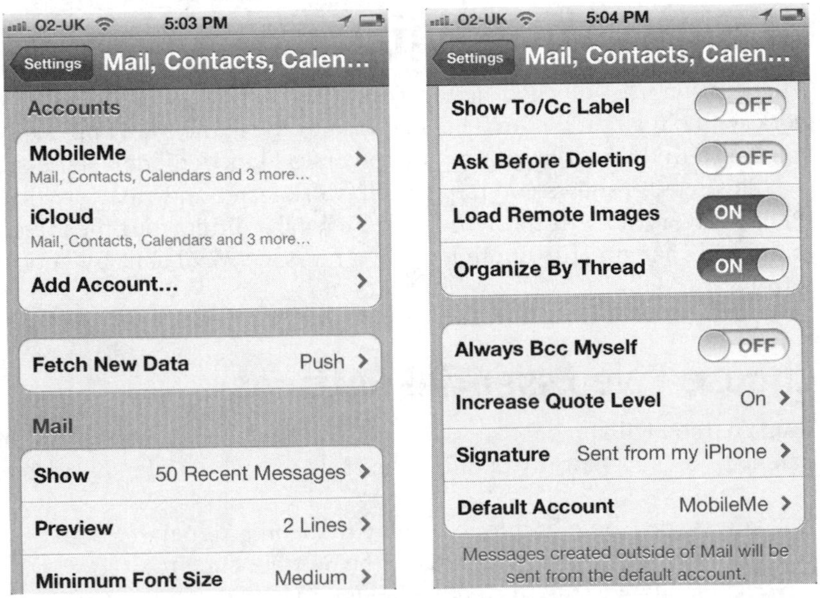

FIGURE 3-23   Open the Mail, Contacts, Calendars screen (left and right) to choose essential settings for the Mail app.

- **Preview**   To choose how much of a preview of each message Mail displays, tap this button, and then tap the appropriate button on the Preview screen: None, 1 Line, 2 Lines, 3 Lines, 4 Lines, or 5 Lines. The more lines you display, the better you can identify each message from the preview—but the fewer previews you can see on the screen at once. Choose your poison.

 If you find Mail is slow to load, try turning off previews by tapping the None button on the Preview screen. See if this change makes an improvement you want to keep.

Set Your Default Account

If you set up two or more e-mail accounts on your iPhone, you need to tell Mail which is the default account to use. To specify the default account, follow these steps:

1. Press the Home button to display the Home screen.
2. Tap the Settings icon to display the Settings screen.
3. Scroll down to the third box, the one that starts with the General button.
4. Tap the Mail, Contacts, Calendars button to display the Mail, Contacts, Calendars screen (shown on the left in Figure 3-24).

.ıl. 3 📶	12:05 PM		.ıl. 3 📶	12:05 PM	
Settings **Mail, Contacts, Calen...**			**Mail...** **Default Account**		
Minimum Font Size	Medium >		mac.com		
Show To/Cc Label	OFF		**Work iCloud Account**	✓	
Ask Before Deleting	OFF		**Home iCloud Account**		
Load Remote Images	ON		Gmail		
Organize By Thread	ON		.Mac Account		
Always Bcc Myself	OFF				
Increase Quote Level	On >				
Signature	>				
Default Account	Work iCloud... >				

FIGURE 3-24   On the Mail, Contacts, Calendars screen, tap the Default Account button to display the Default Account screen (right), and then tap the account you want to use as the default.

5. Tap the Default Account button to display the Default Account screen (shown on the right in Figure 3-24).
6. Tap the account you want to make the default.
7. Tap the Mail, Contacts, Calendars button to return to the Mail, Contacts, Calendars screen.

Get Your Messages Pushed to Your Phone

To get your messages as quickly as possible, set your e-mail accounts on your iPhone to use push. Using push tells the server to "push" out a message to your iPhone as soon as the message arrives at the server, instead of leaving the message at the server until the Mail app checks in for mail.

 Not all e-mail providers support push. If your e-mail account doesn't offer push, you can set your iPhone to check for messages at short intervals instead. This is called *fetch*. And whether you use push or fetch, you can check for messages manually at any point by tapping the Refresh button, the clockwise curling arrow in the lower-left corner of the Mail screen.

To set your iPhone to use push, follow these steps:

1. Tap the Fetch New Data button on the Mail, Contacts, Calendars screen (shown on the left in Figure 3-25) to display the Fetch New Data screen (shown on the right in Figure 3-25).

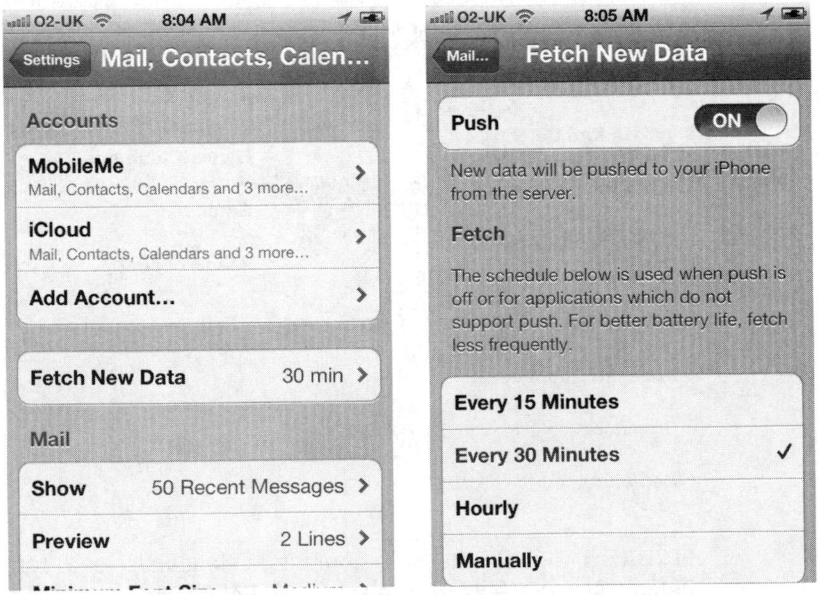

FIGURE 3-25 On the Mail, Contacts, Calendars screen (left), tap the Fetch New Data button to display the Fetch New Data screen (right), and then move the Push switch to the On position.

2. Make sure the Push switch is set to the On position.
3. In the Fetch area, tap to place a check mark on the button for the fetch timing you want to use when push isn't available: Every 15 Minutes, Every 30 Minutes, Hourly, or Manually.
4. Tap the Mail, Contacts, Calendars button to return to the Mail, Contacts, Calendars screen.

Protect Yourself from Images in Spam Messages

These days, it's hard to avoid receiving at least some spam—unwanted messages. When you do, you can simply delete them. But malefactors have another trick up their sleeve: remote images, which are also called *web bugs*. By including in a message a reference to an image stored on a remote server, a spammer can learn not only when you open that message but also your IP address and approximate geographical location.

To avoid this problem, you can set the Load Remote Images switch on the Mail, Contacts, Calendars screen to the Off position. This tells Mail not to load remote images. The images then appear as placeholders in your messages. You can tap a placeholder to display its image—preferably after checking that the message is wholesome.

Change Your Signature

Instead of typing a closing line and your name at the end of each message, you can have Mail add a signature for you automatically. Adding a signature can save you plenty of time and typing, especially if you need to include your business name or contact information.

Your iPhone comes with the default signature of "Sent from my iPhone." This is cute for the first message or two, but you'll likely want to change it before you use your iPhone extensively.

To change your signature, follow these steps:

1. Open the Mail, Contacts, Calendars screen as discussed earlier in this section.
2. Scroll down to the second box of settings under the Mail heading.
3. Tap the Signature button to display the Signature screen (shown here).
4. If there's an existing signature you need to get rid of, tap the Clear button.
5. Type the signature you want.
6. Tap the Mail, Contacts, Calendars button to return to the Mail, Contacts, Calendars screen.

DOUBLE GEEKERY

Create Multiple Signatures by Using Text Shortcuts

At this writing, your iPhone lets you create only a single signature, which it applies to all your accounts. That means any new message you create gets the same signature. Any message you forward or reply to receives the signature too.

This works fine for some people, but if you need to be able to apply different signatures to different messages, you must take another approach.

Instead of creating a signature, go to the Signature screen and tap the Clear button to wipe out whatever signature is there. Then open the General settings screen, tap the Keyboard button to display the Keyboard screen, and set up a text shortcut for each signature or partial signature you want to be able to enter quickly. For example, create a text shortcut for your name, another for your job title, a third for your company name, and a fourth for your address. See the sidebar "Eight Ways to Speed Up Your Typing on the Onscreen Keyboard" earlier in this chapter for instructions on creating text shortcuts.

Once you've created your shortcuts, you can finish an e-mail message quickly with the appropriate signature by typing each shortcut needed and tapping the spacebar.

Batch-Edit Your E-Mail Messages

Instead of dealing with the e-mail messages in an inbox or folder one by one, you can use batch-editing to manipulate multiple messages at once. To use batch-editing, follow these steps:

1. Open the inbox or folder that contains the messages. The left screen in Figure 3-26 shows an example using Gmail.
2. Tap the Edit button in the upper-right corner of the screen to turn on editing mode.
3. Tap the selection button for each message you want to affect, as shown in the right screen in Figure 3-26.
4. Tap the appropriate command button. For example, tap the Move button to display the Mailboxes screen, and then tap the mailbox to which you want to move the messages.

FIGURE 3-26 Tap the Edit button in the upper-right corner of an inbox or folder (left) to turn on Edit mode. You can then tap the selection button for each message you want to affect (right), and then tap the appropriate command button.

Send Your Electronic Business Card to Your Contacts

When you need to share your contact information with someone, send it as an electronic business card attached to an e-mail message. The recipient can then import the data straight into her address book or contact-management program without having to retype any of it.

 You can also send your electronic business card as an attachment to an instant message. Just tap the Message button in the Share Contact Using dialog box, and then address and send the instant message.

To send your electronic business card, follow these steps:

1. Press the Home button to display the Home screen.
2. Tap the Phone icon to display the Phone app.
3. Tap the Contacts button at the bottom of the screen to display the Contacts list.
4. Tap the contact record that contains the data you want to share.

5. Tap the Share Contact button (you may need to scroll down to see it). The Share Contact Using dialog box opens, as shown here.
6. Tap the Email button. Your iPhone causes Mail to start a new message with the contact record attached.
7. Address the message, give it a title and any explanatory text needed, and then tap the Send button.

See Where a Link in a Message Leads

As you know, a link in an e-mail message can display a different address than it actually goes to.

To see the URL to which a link in an e-mail message points, tap and hold the link until Mail displays the Actions dialog box (shown here) with a button for each action you can take with the link. The URL appears at the top. You can then tap the Open button if it's safe to open the link, tap the Copy button if you want to store the URL or share it with someone, or the Cancel button to stop opening it.

Mark a Message as Unread or Flag It as Important

When you receive a new message, Mail places a blue dot to its left in your inbox so that you can see at a glance that it's unread. When you open the message, Mail marks the message as read, and removes the blue dot.

 Tap the iPhone's status bar at the top of the screen to scroll quickly to the top of the open message.

When you're triaging your e-mail, you may want to look quickly at a message but then mark it as unread so that you can see it still needs your attention. To mark the message as unread, tap the Mark button to the right of the message's date (see the left screen in Figure 3-27), and then tap the Mark As Unread button in the dialog box that opens (shown on the right in Figure 3-27).

From the dialog box that opens when you tap the Mark button, you can also tap the Flag button to mark the message with a flag. Mail then displays a flag icon to the left of the message in the inbox or folder.

The flag persists until you remove it by tapping the Mark button and then tapping the Unflag button. You can use flagging for whatever purpose you choose, but its basic advantage over marking the message as unread is that the flag stays in place when you open the message for reading.

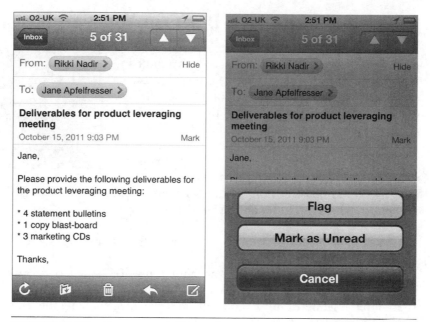

FIGURE 3-27   You can quickly mark a message as unread by tapping the Mark button (left) and then tapping the Mark As Unread button in the dialog box that opens (right).

DOUBLE GEEKERY

Prevent Gmail from Archiving Messages You Want to Delete

Gmail's limitless storage is a huge asset, but if you're like me, you'll want to delete some of your old messages rather than archiving them until doomsday. To stop Gmail from archiving messages, follow these steps:

1. Open the Settings app and display the Mail, Contacts, Calendars screen.
2. In the Accounts list, tap your Gmail account to display its control screen.
3. Tap the Archive Messages switch and move it to the Off position.
4. Tap the Mail, Contacts, Calendars button to return to the Mail, Contacts, Calendars screen.

After you do this, Mail displays the Delete button for your Gmail account instead of the Archive button, and you can delete messages instead of archiving them.

Save a Message as a Draft So You Can Finish It Later

When you don't have time to complete an e-mail message you've started writing, save it as a draft so that you can finish it later. Tap the Cancel button, and then tap the Save Draft button in the Draft dialog box (shown here) that Mail displays. Mail saves the message in your Drafts folder. (If the Drafts folder doesn't yet exist, Mail creates it.)

You can quickly reopen the last draft message you were working on by simply tapping and holding the Compose button in Mail. To open an older draft message, go to the Drafts mailbox for the account, and then tap the message:

1. Tap the Mailboxes button in the upper-left corner of the screen to display the Mailboxes screen.
2. In the Accounts box, tap the account you were using when you created the draft. The list of folders for that account appears.
3. Tap the Drafts button. The Drafts folder opens, showing the list of draft messages.
4. Tap the message you want to open.

Change the Account You're Sending a Message From

If you start an e-mail message from the wrong account, you don't need to scrap the message and start again. Just tap the From field to expand the Cc/Bcc and From area, tap the From field again, and then tap the address you want to use on the spin wheel that appears (see the illustration).

Apply Formatting to Text in a Message

If you want to make parts of a message you're composing stand out, you can apply boldface, italics, or underline (or two of those three, or all three).

To apply formatting, follow these steps:

1. Select the text you want to format.
2. To display the next section of the bar (shown on the left in the illustration), tap the > button on the bar that appears.

3. Tap the BIU button to display the Bold, Italics, Underline bar (shown on the right in the preceding illustration).

4. Tap the Bold button, the Italics button, or the Underline button, as needed.

Change the Quote Level in a Message

Another way to make text in a message you're writing stand out is to mark it as being indented. You can do this by selecting the text, tapping the > button on the bar that appears, tapping the Quote Level button, and then tapping the Increase button on the bar that appears (as shown here). You can also tap the Decrease button to decrease the indentation of text that's already indented.

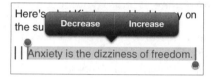

Send a Message to a Group Without Revealing the E-Mail Addresses

When you need to send an e-mail message to a group of people who don't necessarily know each other, don't put all the e-mail addresses in the To box or the Cc box, because each recipient will be able to see all the other addresses.

Instead, put your own address in the To box, and then tap the Cc/Bcc, From area to display the Bcc field. Put each e-mail address in this field, and each recipient will see only his or her own e-mail address, not those of the other Bcc recipients. (They'll also see your e-mail address, both in the To field and in the Sender field.)

Project 25: Give Presentations Straight from Your iPhone

If you travel for business, chances are you need to give presentations. If you have a laptop with you, great, because that's still the best tool for giving presentations. But if you have only your iPhone, don't worry—you can give a fine presentation using it. You'll just need to do a bit more preparation and arm yourself with the right apps and cables.

In this section, we'll first run through your options for giving a presentation from your iPhone, and then we'll look at how to use each of those options.

Choose How You Will Give Your Presentation

First, choose how you will give your presentation. You have three main options:

- **Connect your iPhone physically to a projector, monitor, or TV** This approach is just like using a laptop and works well for standard presentation situations—for example, presenting to a group of people who are in the same room and looking at the same screen. The best app for giving the presentation is Apple's Keynote, which you can also use for creating and editing presentations.
- **Connect your iPhone wirelessly to one or more laptops or desktops** This approach requires no cable and enables you to send the presentation wirelessly to a web browser on one or more computers within broadcasting distance. You can't use Keynote on your iPhone for this—you need to use a third-party app instead. Your presentation is limited to PDF files and photos.
- **Use your iPhone as a remote control for Keynote on the Mac** With this approach, you're using the iPhone to control the presentation, but the presentation is actually running on a Mac that's connected to a projector, monitor, or TV.

We'll cover these possibilities in turn in the following sections.

Give a Presentation Using Keynote and a Projector, Monitor, or TV

In this section, we'll look at how to give a presentation using Apple's Keynote app on your iPhone. You'll need to connect your iPhone to the projector, monitor, or TV on which you will show the presentation.

Add Keynote to Your iPhone

The first step is to add Keynote to your iPhone if you don't have it already. Go to the App Store using either your iPhone or iTunes on your computer, buy Keynote (it costs $9.99), and then download and install it.

Prepare Your Presentation

Next, prepare your presentation. Normally, you'll want to use one of these ways:

- **Create the presentation in Keynote on the Mac** When the presentation is ready for use, you can transfer it to your iPhone by using iTunes' File Sharing feature.
- **Create the presentation in PowerPoint on Windows or the Mac** In this case, too, you can transfer the presentation to your iPhone by using iTunes' File Sharing feature.

- **Create the presentation in Keynote on your iPhone**   Keynote on the iPhone offers surprisingly good features for creating presentations on the small screen. To start a new presentation, tap the New (+) button in the upper-left corner of the screen, tap the Create Presentation button in the pop-up panel (shown here), and then tap the theme you want on the Choose A Theme screen (see Figure 3-28).

FIGURE 3-28   On the Choose A Theme screen, tap the theme you want to base the new presentation on.

DOUBLE GEEKERY

Double-Check Any Presentation You Import into Keynote for iPhone

Always check a presentation closely after you import it into Keynote for iPhone. While Keynote supports as many features the iWork team has been able to pack in, it doesn't support all the features of its older sibling, Keynote for Mac, let alone all the features of PowerPoint.

Here are three examples:

- **Fonts**   When Keynote doesn't have a font that the presentation uses, Keynote substitutes a similar font. Unless you're heavily into design (or your audience is), this substitution usually makes little difference.
- **3D charts**   Keynote for iPhone doesn't support 3D charts, so it converts them to 2D charts.
- **Build order**   Keynote may change the build order of objects on some slides. This can cause some entertaining surprises.

So after you import a presentation, go through it, and make sure that each slide looks okay.

Connect Your iPhone to a Projector, Monitor, or TV

If you're going to give a conventional presentation directly from your iPhone, you need to connect it to a projector, a monitor, or a TV. (We'll ignore the possibility of showing the presentation to a tiny audience on your iPhone's screen—you know how to do this.)

To connect your iPhone to a projector, monitor, or TV, you need the right kind of cable. These are the three types of cables you're most likely to need:

- **Apple Digital AV Adapter**   This short cable has a Dock Connector on one end and an HDMI port and a Dock Connector port on the other end. You plug the Dock Connector into your iPhone, and then plug an HDMI cable into the other end of the cable and into your TV. You can connect your iPhone's USB cable to the Dock Connector port to charge the iPhone.

 If you have to choose among HDMI, VGA, and composite, choose HDMI every time, because it will give you higher quality.

- **Apple VGA Adapter**   This short cable has a Dock Connector at the iPhone's end and a female VGA connector at the other end to which you connect a standard VGA cable running from the projector or monitor.
- **Apple Composite AV Adapter**   This cable has a Dock Connector on one end and three RCA plugs on the other end—red and white connectors for the audio channels, and a yellow connector for the video. The RCA end also has a USB cable for powering the iPhone. Use this cable to connect to a TV that has composite video jacks.

After you connect your iPhone to the output device, Keynote appears mirrored on it. So when you change the slide on your iPhone, the new slide appears on the output device as well.

Give Your Presentation on the iPhone

To give your presentation on the iPhone, open the presentation in Keynote, display the first slide, and then tap the Play button.

The presentation starts playing on both your iPhone's screen and on the screen you've connected to your iPhone. You can display the next slide by tapping anywhere on the screen or by swiping a finger from right to left across the screen. If you need to display the previous slide, swipe from left to right across the screen.

To end the presentation, pinch inwards with two fingers on the screen.

Connect Your iPhone Wirelessly to One or More Computers

When you can't use Keynote or establish a physical connection between your iPhone and a projector, monitor, or TV, you can give a presentation wirelessly. Connecting wirelessly also enables you to give your presentation on multiple screens at once, which can be useful for lab or classroom situations.

At this writing, the best bet for presenting wirelessly like this is AirProjector, which you can get from the App Store for $2.99. Start by trying the free version, AirProjector Free, to see how it works for you, and then move to the full version if you need to.

AirProjector enables you to broadcast PDF files or photos from your iPhone to a web browser on a laptop or desktop computer. You can either use that computer's screen or connect that computer to a projector for a large-screen presentation. On the computer, you simply enter in the web browser the IP address and port number that AirProjector is using on the iPhone. The browser then displays the photo or PDF file you display on the iPhone's screen.

Use Your iPhone as a Remote Control for Your Mac

If you're giving a presentation from a Mac, you can use your iPhone as a remote control.

To do so, download and install the Keynote Remote app (which costs $0.99) from the iTunes Store. Search for "Keynote Remote," and you'll find it in moments.

Once you've installed the app, locate it—you'll find it appears as "Remote" rather than "Keynote Remote"—and then run it. You'll see first a screen telling you that you haven't linked it to Keynote, as shown on the left in the next illustration. Tap the Link

To Keynote button. Keynote Remote then automatically displays the Settings screen, shown on the right in the illustration.

To set up the link to Keynote, follow these steps:

1. Tap the New Keynote Link button on the Settings screen. Keynote Remote displays the New Link screen, which contains a fresh passcode for linking to Keynote.
2. Launch Keynote on your Mac, or switch to Keynote if it's already running.
3. In Keynote, choose Keynote | Preferences to display the Preferences window.
4. Click the Remote tab to display its contents, as shown here. Your iPhone should appear with a Link button to its right.

5. Make sure the Enable iPhone And iPod touch Remotes check box is selected.
6. Click your iPhone's Link button to display the Add Remote For iPhone And iPod touch dialog box (shown here).

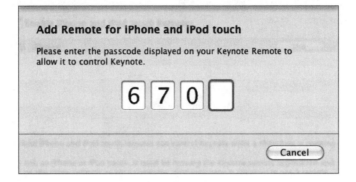

7. Type the passcode that the Remote app is displaying. Keynote checks the passcode and then closes the Add Remote For iPhone And iPod touch dialog box automatically. Your iPhone then appears with an Unlink button on the Remote tab of the Preferences window
8. Click the Close button (the red button) to close the Preferences window.

4 Security and Troubleshooting Geekery

In this chapter, we'll look at how to secure your iPhone against theft or intrusion, how to track it down if you lose it, and how to wipe the data from your iPhone if you can't recover it.

I'll also show you how to deal with problems closer to home. We'll consider your options for using your iPhone safely in wet or dirty conditions, troubleshoot software and hardware problems, and restore your iPhone to factory settings if its software becomes messed up or you need to sell it.

Project 26: Secure Your iPhone Against Theft or Intrusion

Packed with not only the highest technology around but also your priceless personal secrets and business intelligence, your iPhone is a tempting target for thieves, who know they can readily sell it or the data it contains for a fistful of dollars. So no matter how firmly you grip your iPhone in public or how well you hide it at home, you need to secure it effectively in case it goes missing (presumed stolen).

Locking your iPhone in a fire- and waterproof strongbox in your bank might keep it physically safe, but it wouldn't be much use to you. Given that you need to keep your iPhone with you all the time, securing your iPhone consists of preventing other people from accessing the data on it.

There are two main ways to secure the data on your iPhone:

- Set your iPhone to lock itself shortly after you stop using it.
- Require a passcode to unlock the iPhone. If necessary, set your iPhone to wipe its data if someone enters the wrong passcode too many times in succession.

 If you're setting up an iPhone for someone else to use, you can apply restrictions to what the user can do with the iPhone. To apply restrictions, choose Settings | General | Restrictions, tap the Enable Restrictions button, type in a passcode to protect the restrictions, and then choose the details on the Restrictions screen. For example, you can prevent the user from installing apps, limit them to age-appropriate movies and apps, and stop them from adding friends in Game Center.

Set Your iPhone to Lock Itself Automatically

First, set your iPhone's Auto-Lock feature to lock the iPhone automatically a short time after you stop using it.

To set up Auto-Lock, follow these steps:

1. Press the Home button to display the Home screen.
2. Tap the Settings icon to display the Settings screen.
3. Scroll down to the third box, and then tap the General button to display the General screen.
4. Scroll down to the bottom of the screen, and then tap the Auto-Lock button to display the Auto-Lock screen (shown in Figure 4-1).
5. Tap the button for the interval you want: 1 Minute, 2 Minutes, 3 Minutes, 4 Minutes, 5 Minutes, or Never. The shorter the interval, the safer, so try the 1 Minute setting and see how well it works for you.
6. Tap the General button to return to the General screen.

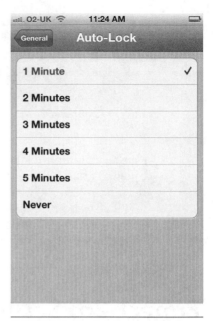

FIGURE 4-1 On the Auto-Lock screen, choose as short an interval as is practical for the way you use your iPhone.

 You can also lock your iPhone at any point by pressing the Sleep/Wake button when the iPhone is unlocked. To make your iPhone lock the moment you put it to sleep, set the Require Passcode setting (discussed later in this chapter) to Immediately.

Protect Your iPhone with a Passcode Lock

Next, protect your iPhone with a passcode lock. The passcode is a sequence of characters that you must type each time you unlock the iPhone from the lock screen. The nearby sidebar "Choose Between a Simple Passcode and a Complex Passcode" explains the ins and outs of passcodes.

 If your company or organization provides your iPhone, an administrator may apply a configuration profile that compels you to use a passcode on the iPhone. If you find you cannot change the passcode settings on your iPhone, you will know that a profile is installed.

DOUBLE GEEKERY

Choose Between a Simple Passcode and a Complex Passcode

You can protect your iPhone with either a simple passcode or a complex passcode:

- **Simple passcode**   Four digits—for example, 1924. This is the default type, and it works well for general-purpose needs.
- **Complex passcode**   A variable number of characters that mixes letters and other characters with digits.

A complex passcode can provide much greater security than a simple passcode:

- **You can set a longer passcode**   A longer passcode is harder to crack because it contains more characters. This is true even if the passcode consists only of numbers rather than letters and non-alphanumeric characters.
- **You can include letters**   Including letters as well as numbers greatly increases the strength of the passcode even at short lengths.
- **You can include non-alphanumeric characters**   Including non-alphanumeric characters (such as symbols—&*#$!, and so on) increases the strength of the passcode even further.

The Enter Passcode screen that prompts you for the passcodes shows whether the iPhone is using a simple passcode or a complex passcode. For a simple passcode, the Enter Passcode screen displays four boxes and a numeric keypad, as shown on the left in the next illustration. For a complex passcode, the Enter Passcode screen displays a text box and the QWERTY keyboard, as shown on the right in the illustration.

Whether you should use a simple passcode or a complex passcode depends on how much security you feel you need. Having to enter a long passcode for each unlock can make it hard to jot down quick notes and use the iPhone to its full potential.

Keep these points in mind when deciding which type of passcode to use:

- **A simple passcode may be strong enough with auto-erase** Given enough time and tries, anyone can break a simple passcode by plodding through all 10,000 possible numbers until they hit the jackpot. Your iPhone makes this harder by automatically disabling itself for increasing periods of time—1 minute, 5 minutes, 15 minutes, 60 minutes, and so on—as the wrong passcodes hit in sequence (see the next illustration). A determined attacker can keep entering passcodes as soon as the iPhone starts accepting them again, but if you set your iPhone to erase its data automatically after a handful of failed attempts to enter the passcode, your data should be pretty safe—unless you've chosen a personal number that the attacker can guess (for example, your birth year, which is a regrettably popular passcode).

- **With a complex passcode, you may not need auto-erase** If you use a complex passcode of a certain length (say eight or more characters) and including both alphanumeric and non-alphanumeric characters, you may consider it strong enough that your iPhone doesn't need auto-erase. But if your iPhone's contents are highly valuable or important, you'll probably want to use auto-erase.
- **A complex passcode can be shorter than a simple passcode** Because the Enter Passcode screen for a complex passcode gives no indication of the passcode's length, you may be able to bluff an attacker by setting a short, letters-only passcode (for example, aq) rather than a lengthy excerpt from a monkey's attempts to type *Hamlet*. A short passcode like this is easy for you to remember and type, so you can set a low number for the Maximum Number Of Failed Attempts setting as a safety net.

To set your passcode lock and (if you want) automatic wiping, follow these steps:

1. Choose Home | Settings | General to open the General screen in Settings.
2. Scroll down until you see the box starting with the Auto-Lock button. The left screen in Figure 4-2 shows this section of the General screen.
3. Tap the Passcode Lock button to display the Passcode Lock screen (shown on the right in Figure 4-2).
4. If you want to use a simple passcode—a four-digit number—make sure the Simple Passcode switch is set to the On position. If you want to lock your iPhone down more tightly by using a complex passcode, move the Simple Passcode switch to the Off position.
5. Tap the Turn Passcode On button to display the Set Passcode screen. For a simple passcode, you'll see the Set Passcode screen shown on the left in Figure 4-3; for a complex passcode, you'll see the Set Passcode screen shown on the right in Figure 4-3.
6. Tap the numbers or characters for the passcode:
   - **Simple passcode**   When you've entered four numbers, your iPhone displays the Set Passcode: Re-enter Your Passcode screen automatically.
   - **Complex passcode**   Tap the .?123 button when you need to reach the keyboard with numbers and some symbols. From here, you can tap the # + = button to reach the remaining symbols, punctuation characters, and currency characters. When you've finished entering the passcode, tap the Next button to display the Set Passcode: Re-enter Your Passcode screen.

FIGURE 4-2   On the General screen (left), tap the Passcode Lock button to display the Passcode Lock screen (right).

FIGURE 4-3 On the Set Passcode screen, enter either a simple four-digit passcode (left) or a complex passcode as long—or as short—as you like (right).

7. Tap the numbers or characters for the passcode again; for a complex passcode, tap the Done button when you finish. Your iPhone displays the Passcode Lock screen again. This time, all the options are enabled, as shown in the screen on the left in Figure 4-4.
8. Look at the Require Passcode button to see how quickly the passcode requirement kicks in: Immediately, After 1 Minute, After 5 Minutes, After 15 Minutes, After 1 Hour, or After 4 Hours. If you need to change the setting, follow these steps:
    a. Tap the Require Passcode button to display the Require Passcode screen (shown on the right in Figure 4-4).
    b. Tap the button for the interval you want.
    c. Tap the Passcode Lock button to return to the Passcode Lock screen.

 For the Require Passcode setting, the Immediately option is by far the most secure, because it locks your iPhone the moment you put it to sleep or the Auto-Lock feature runs. But if you tend to put your iPhone to sleep and then immediately think of something else you must note, you may find the After 1 Minute option a better choice, because it will allow you to unlock your iPhone to jot down your new item.

FIGURE 4-4 After you set a passcode, the remaining options on the Passcode Lock screen become available (left). Tap the Require Passcode button to display the Require Passcode screen (right), on which you can set the interval after which your iPhone requires the passcode.

9. Move the Siri switch on the Passcode Lock screen to the Off position to prevent Siri from running when your iPhone is locked.

To protect your iPhone and yourself, you must prevent Siri from bypassing the lock screen. Otherwise, anyone who can speak intelligibly to your iPhone can take a wide range of actions on your behalf—from harmless actions such as browsing the Web to potentially harmful actions such as placing phone calls and sending instant messages and e-mail messages.

10. If you want your iPhone to erase its contents after ten failed attempts to crack the passcode, tap the Erase Data switch and move it to the On position. Then tap the Enable button in the confirmation dialog box (shown here) that appears.

DOUBLE GEEKERY

Don't Let the Police Connect Your iPhone to a UFED

Your iPhone has strong security against conventional threats—but watch out for the police.

Various police departments use devices called Universal Forensic Extraction Devices, or UFEDs. If you allow the police to connect your iPhone to a UFED, the UFED can grab all of the data from your iPhone, even if you have secured your iPhone with a password. Most UFEDs use a physical connection to the Dock Connector, but some models have Bluetooth capability as well.

The American Civil Liberties Union is arguing that this data extraction constitutes an unreasonable search under the Fourth Amendment—but at this writing, the argument is still unresolved.

If you have the choice (and you may not have), don't let the police connect your iPhone to a UFED, even if you've secured your iPhone with a passcode and automatic wiping.

Project 27: Use Your iPhone Safely in Wet or Dirty Conditions

To stay in touch with your contacts and in control of your life, you'll probably want to take your iPhone with you more or less everywhere. Chances are that'll include plenty of places that are wet, dirty, or both.

Your iPhone dreads water even more than it fears gravity and small children, and it's not big on dirt either—so you'll need to protect it. For most people, that means using a case.

You can find a wide variety of cases at bricks-and-mortar stores (such as the Apple Store or Best Buy) and a truly terrifying variety online. If you don't find what you need among the loads of cases at major sites such as Amazon and eBay, either search the Internet or visit case manufacturers such as OtterBox (www.otterbox .com), Speck Products (www.speckproducts.com), Marware (www.marware.com), RadTech (www.radtech.us), or DecalGirl (www.decalgirl.com).

If you need to keep your iPhone dry, the first question is whether you need the case to be water-resistant or actually waterproof. You can find plenty of protective cases that are water-resistant enough for general use but that make your iPhone's ports and buttons easily accessible. For example, many protective cases use rubber plugs to close cutouts for the headphone jack, camera lens, mute switch, and Dock Connector port. You can easily swing out a rubber plug when you need to use the port, switch, or lens; but when the plug is in place, it keeps rain, water splashes, or dust out of the iPhone. But this arrangement is only water-resistant—it's not waterproof.

If you actually need to be able to drop your iPhone in water without major sadness and expense ensuing, you need the next step—either a fully waterproof case or a bag or box you can put the iPhone in to keep the water out.

Here are three sources of waterproof iPhone cases:

- **Amazon.com**   At this writing, Amazon offers various waterproof cases including the Keystone ECO Waterproof Case (around $39.99), PaleKai (around $45), and the Waterproof Case for Apple iPhone 4/4S ($99.99 list, but usually available for much less).

 Read the buyer reviews on Amazon.com to get a clearer idea of what a particular case is good for, how well it delivers on its promises, and what its weak points are.

- **Waterproof iPhone Case**   As its name suggests, this site (www .waterproofiphonecase.net/) specializes in waterproof iPhone cases such as the Magellan Waterproof iPhone Tough Case ($79.99; includes a built-in battery and enhanced GPS) and the Grace Digital Audio Eco Extreme Waterproof iPhone Case (around $45; includes a built-in speaker).
- **eBay**   As you know, you can find just about anything on eBay—and that includes plenty of cases that claim to be waterproof. You can find both high-end cases, such as the LifeProof Case for iPhone 4/4S (around $80; look also on www.lifeproof.com), and low-end cases.

 DOUBLE GEEKERY

Know What the IPX Certifications Mean

When you're shopping for waterproof cases, you'll see certification numbers such as IPX7 and IPX8. Here, IP stands for *ingress protection*—how much protection the case provides against stuff getting in. The following list shows what the IPX ratings mean for liquid ingress protection.

IPX Rating	Protection Against
1	Vertically dripping water
2	Water dripping at an angle of up to 15 degrees
3	Water spraying at an angle of up to 60 degrees
4	Water splashing from any direction
5	Water jets from any direction
6	Powerful water jets
7	Immersion up to 1 meter deep
8	Immersion of more than 1 meter deep

So if you want your iPhone to survive a drop into household water, IPX7 has you covered. If you're planning to take your iPhone swimming or diving, you'll want IPX8, which typically means the case is hermetically sealed.

Even a teaspoon of water in the wrong place can ruin your iPhone, so you'll want to be sure you can trust the case you buy. This is one area where saving money by buying a no-name brand can cost you dear, so you may decide that you want to stick with a big-name brand—perhaps one that provides a guarantee.

Fully waterproof cases are great if that's what you need, but because they're sealed, they tend to make access to the iPhone's ports difficult. You can use the screen as usual through the case, and play music wirelessly by using AirPlay, but you'll typically need to remove all or part of the case in order to recharge the iPhone. The bigger waterproof cases simply snap open, but those that fit more snugly can take time and effort to remove.

If you need to make your iPhone fully waterproof only on special occasions, you may prefer the next type of waterproofing: instead of getting a waterproof case, get a waterproof bag or box into which you can put your iPhone in its existing case (if any). If you need to use your iPhone, you'll have to take it out of the case. But the priority here is keeping the iPhone dry and happy.

DOUBLE GEEKERY

The Easy Way to Check How Wet Your iPhone Just Got

Many an iPhone has taken a dive into a toilet bowl—most of them accidentally, but some of them deposited there by helpful toddlers (see YouTube for evidence of this).

Apple is well aware of your iPhone's propensity for a triple backward somersault followed by an otter-smooth entrance into the water. Your iPhone's warranty doesn't cover liquid damage, and Apple has built in a way of checking instantly whether you've soaked it.

Your iPhone contains two Liquid Contact Indicators to enable a technician to tell whether water has gone into the iPhone. One of the Liquid Contact Indicators is in the headphone port. The other is in the Dock Connector port. The Liquid Contact Indicators turn red when the iPhone gets wet.

If you think your iPhone may have gotten wet—or (let's be honest) if your iPhone has gotten wet and you want to find out how badly—shine a light into first the headphone port and then the Dock Connector port. If you see red, you'll know that Apple won't be fixing or replacing your iPhone.

Even so, all may not be lost. Try leaving your iPhone to dry for about three days, either in a warm (but not hot) and well-ventilated place, or snuggled into a nest of desiccant packets. If you can't get desiccant packets quickly, fill a sock with dry rice and wrap your iPhone in the sock. When your iPhone has dried out thoroughly, cross your fingers and try turning it on.

You can get custom waterproof cases that are just the right size for an iPhone and its case. For example, the OtterBox 2000 (around $17.95; www.otterbox.com) is a waterproof box that's big enough for an iPhone in any case up to the size of the OtterBox Defender case. (Most iPhone cases are smaller than the Defender, so most will fit inside the OtterBox 2000—but check if you're in doubt.) If you need to haul more gear around, get one of the larger OtterBox cases that'll take not just your iPhone but also your camera or other gear you need to keep dry.

If your needs are more modest, experiment with sealable plastic bags and sandwich boxes. Either works well in a pinch—and you probably have enough of both in the kitchen to keep your iPhone safe and dry without spending a cent.

Project 28: Troubleshoot Software and Hardware Problems

Apple has made your iPhone and its operating system, iOS, as stable and reliable as possible. But even so, you may run into software and hardware problems now and then.

This project shows you five essential moves:

- **Force quit an app**   When an app stops responding, you can force it to quit.
- **Restart your iPhone**   Restarting your iPhone can clear up software and hardware problems.
- **Hardware reset**   When a restart doesn't do the trick, you can perform a hardware reset. This is essentially a restart on steroids. It doesn't affect the data or settings on the iPhone.
- **Software reset**   The next stage is to reset all the settings on your iPhone. This move loses your custom settings but doesn't affect your data.
- **Erase all content and settings**   If the software reset doesn't clear the problems, you can erase all your content and settings from your iPhone. Before you do this, you need to sync your iPhone or (if it won't sync) save any content that's only on the iPhone. After erasing all content and settings, you sync the content and settings back to the iPhone.

Beyond these five moves, there's the heaviest-duty move: restoring your iPhone to factory settings. I'll show you how to do this in the next project, including how to put your iPhone into Device Firmware Upgrade mode if needed.

Force Quit an App That Has Hung

Normally, the apps you've launched on your iPhone just keep running until you turn your iPhone off.

For example, say you're working in the Mail app, and you press the Home button to display the Home screen so that you can launch another app. iOS doesn't close Mail; instead, Mail keeps running in the background, where you can't see it. When you go back to Mail, either by tapping its icon on the Home screen or by using the quick switching feature, Mail will be as you left it. So if you've left a message half-written, it'll still be there for you to continue.

When you use the Home screen to switch to a different app, iOS keeps the app you were previously using suspended in the background, where you can't see it. When you go back to that app, you'll find it doing what it was doing before.

If an app stops responding, you can close it by "force quitting" it—in other words, forcing it to quit. To force quit a program, follow these steps:

1. Press the Home button twice in rapid succession to display the app-switching bar.
2. If the app you want to force quit doesn't appear on the first screen displayed of the app-switching bar, scroll left or right until you can see it.
3. Tap and hold the app's icon on the app-switching bar until the icons start to jiggle and a Close button (a red circle with a horizontal white bar across it) appears at the upper-left corner of each icon, as shown here.
4. Tap the Close button for the app.
5. Press the Home button to stop the icons jiggling.

Restart Your iPhone

If your iPhone is not running stably, try restarting it. Follow these steps:

1. Hold down the Sleep/Wake button until the screen shows the message Slide To Power Off.
2. Tap the slider and drag it to the right. The iPhone shuts down.
3. Wait a few seconds, and then press the Sleep/Wake button again. Hold the button down for a second or two until the Apple logo appears. The iPhone then starts.

Perform a Hardware Reset

If you're not able to restart your iPhone as described in the previous section, try a hardware reset. Hold down the Sleep/Wake button and the Home button together for around ten seconds until the Apple logo appears on the screen, and then release them. The iPhone then restarts.

Perform a Software Reset

If performing a hardware reset (as described in the previous section) doesn't clear the problem, you may need to perform a software reset. This action resets the iPhone's settings but doesn't erase your data from it.

To perform a software reset, follow these steps:

1. Press the Home button to display the Home screen.
2. Tap the Settings icon to display the Settings screen.
3. Scroll down to the third box, and then tap the General button to display the General screen.
4. Scroll down to the bottom and tap the Reset button to display the Reset screen (shown on the left in Figure 4-5).
5. Tap the Reset All Settings button, and then tap the Reset All Settings button in the confirmation dialog box (shown on the right in Figure 4-5).

Your iPhone then restarts. When it is running again, tap the Settings button on the Home screen to open the Settings app, and start choosing the settings that are most important to you. For example, connect to a wireless network, set the screen brightness, and choose which notifications to receive.

Erase the Content and Settings on Your iPhone

If even the software reset doesn't fix the problem, try erasing all content and settings. Before you do so, remove any content you have created on your iPhone and not yet synced—assuming the iPhone is working well enough for you to do so. For example,

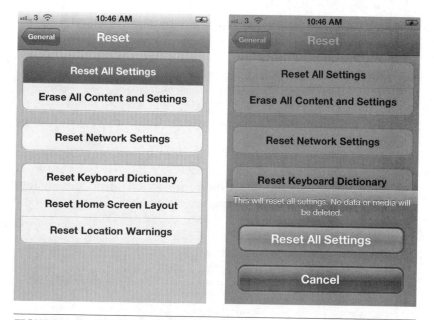

FIGURE 4-5   Tap the Reset All Settings button on the Reset screen (left), and then tap the Reset All Settings button in the confirmation dialog box (right).

send to yourself via e-mail any notes that you have written on the iPhone, or sync the iPhone to your computer to transfer any photos you have taken with its camera.

To erase the content and settings, follow these steps:

1. Press the Home button to display the Home screen, unless you're already there.
2. Tap the Settings icon to display the Settings screen.
3. Scroll down to the third box, and then tap the General button to display the General screen.
4. Tap the Reset item to display the Reset screen.
5. Tap the Erase All Content And Settings button, and then tap the Erase iPhone button on the first confirmation screen.
6. Tap the Erase iPhone button on the second confirmation screen. (Erasure is such a serious move that the iPhone makes you confirm it twice.)

After erasing all content and settings, sync the iPhone to load the content and settings back onto it.

Project 29: Restore Your iPhone to Factory Settings

If your iPhone's software gets really messed up, you may need to restore it to factory settings.

Restoring your iPhone to factory settings wipes out all third-party apps, leaving only the built-in apps—Safari, Mail, Photos, Notes, Camera, and so on. So after restoring to factory settings, you'll need to reload all your third-party apps from your backup on your computer or from the App Store.

To restore the iPhone, follow these steps:

1. Connect the iPhone to your computer, and wait for it to appear in the Source list in iTunes.
2. Click the iPhone's entry in the Source list to display the iPhone control screens.
3. If the Summary screen isn't displayed already, click the Summary tab to display it.
4. Click the Restore button. iTunes displays a confirmation dialog box, as shown here, to make sure you know that you're about to erase all the data from your iPhone.

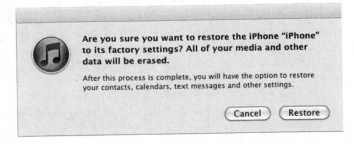

Are you sure you want to restore the iPhone "iPhone" to its factory settings? All of your media and other data will be erased.

After this process is complete, you will have the option to restore your contacts, calendars, text messages and other settings.

Cancel   Restore

If a new version of the iPhone software is available, iTunes prompts you to restore and update your iPhone instead of merely restoring it. Click the Restore And Update button if you want to proceed; otherwise, click the Cancel button.

5. Click the Restore button to close the message box. iTunes wipes the iPhone's contents, and then restores the software, showing you its progress while it works.

6. At the end of the restore process, iTunes restarts the iPhone. iTunes displays an information message box for ten seconds while it does so. Either click the OK button or allow the countdown timer to close the message box automatically.

7. After the iPhone restarts, it appears in the Source list in iTunes. Instead of the iPhone's regular tabbed screens, the Set Up Your iPhone screen appears (see Figure 4-6).

8. To restore your data, make sure the Restore From The Backup Of option button is selected, and verify that the correct iPhone appears in the drop-down list.

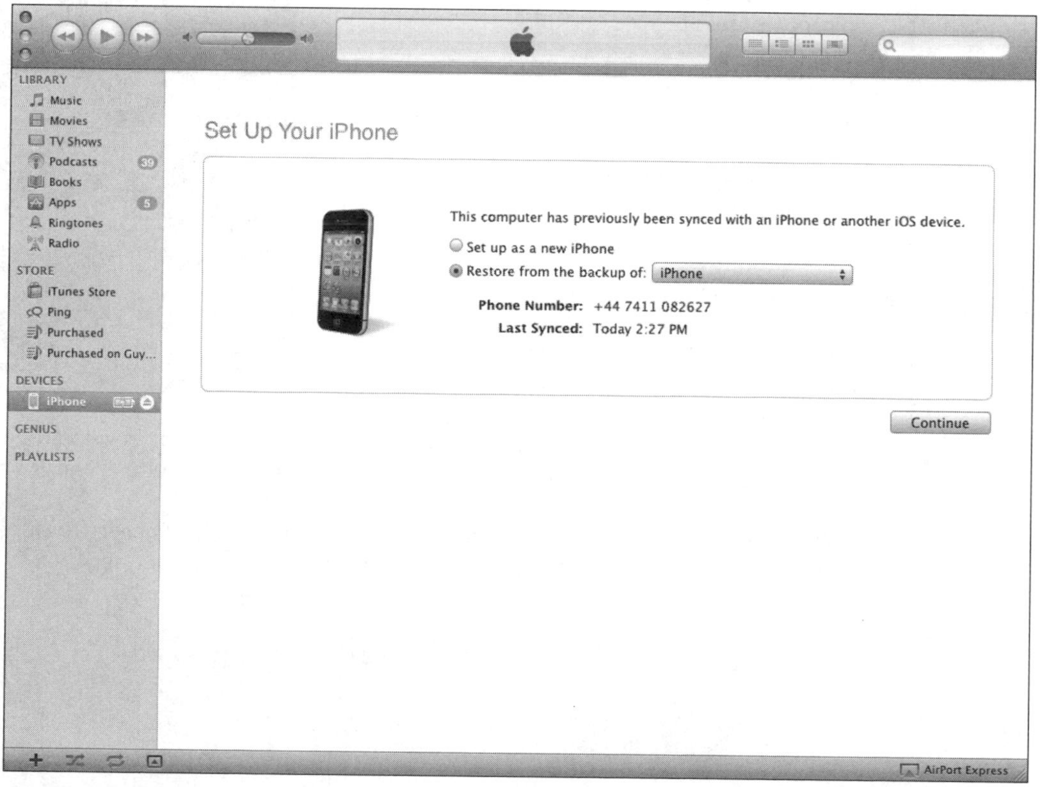

FIGURE 4-6   After restoring the iPhone's system software, you will normally want to restore your data from backup. The alternative is to set up the iPhone as a new iPhone.

9. Click the Continue button. iTunes restores your data and then restarts the iPhone, displaying another countdown message box while it does so. Either click the OK button or allow the countdown timer to close the message box automatically.

10. After the iPhone appears in the Source list in iTunes following the restart, you can use it as normal.

 DOUBLE GEEKERY

Secret Tricks for Recovering from Restore Failures

Sometimes, when you try to restore your iPhone as described in the main text, the restore operation fails in one of these ways:

- The iPhone shows the Connect To iTunes screen (see the illustration), but when you connect the iPhone, it doesn't appear in iTunes.
- Your iPhone keeps restarting, but it doesn't get as far as the Home screen.
- Your iPhone stops responding during the restore operation. The screen may show only the Apple logo or the Apple logo and a progress bar that has stopped moving.

If you run into any of these problems, try using recovery mode. Follow these steps:

1. Disconnect the USB cable from your iPhone.
2. Press and hold the Sleep/Wake button on the top of the iPhone until the Slide To Power Off slider appears, and then tap the slider and drag it to the right. The iPhone powers off.
3. Press and hold the Home button while you plug the USB cable into the iPhone's Dock Connector port. You'll see the iPhone turn on.
4. Keep holding down the Home button until your iPhone shows the Connect To iTunes screen (shown earlier in this sidebar), and then release the Home button.
5. Wait until iTunes displays the Recovery Mode dialog box (shown here).

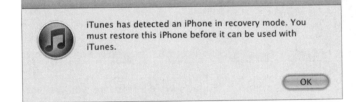

6. Click the OK button (it's the only choice). iTunes displays the Summary tab of the iPhone control screens (as shown here) with only the Restore button enabled.

7. Click the Restore button, and then follow through the process of restoring the iPhone.

Project 30: Track Your iPhone Wherever It May Roam

Your iPhone's Find My iPhone feature enables you to track your iPhone down if you lose it or someone steals it. You can display a message on the iPhone—for example, asking whoever has found the iPhone to call you to arrange its return—or you can wipe the data on the iPhone remotely if you decide you're not going to get it back.

 At this writing, Find My iPhone also works for subscribers to Apple's MobileMe service. But Apple has announced that it will close MobileMe on June 30, 2012, is no longer selling MobileMe subscriptions, and is encouraging MobileMe subscribers to move to iCloud.

To use Find My iPhone, you must have an Apple ID. If you already have an iCloud account, you're all set. If you don't have an Apple ID, you can set one up in just a minute or two.

If you apply restrictions to an iPhone that someone else uses, you can use Find My iPhone to track that iPhone too. For example, you may want to keep tabs on where your child is, or be able to locate an employee (or at least the employee's phone).

Turn On Find My iPhone

To turn on Find My iPhone, follow these steps:

1. Press the Home button to display the Home screen.
2. Tap the Settings icon to display the Settings screen.
3. Scroll down to the third box, the one that starts with the General button.
4. Tap the iCloud button to display the iCloud screen:
   - If you haven't yet set up an iCloud account on your iPhone, you'll see a screen like the left one in Figure 4-7. Type your Apple ID and password, and then tap the Sign In button.

 If you don't have an Apple ID yet, tap the Get A Free Apple ID button at the bottom of the iCloud screen, and then follow through the process for setting up the iCloud account. When you've done so, use the Apple ID to sign in.

   - When you have set up your iCloud account on your iPhone (and signed in), you'll see a screen like the right one in Figure 4-7.
5. Scroll down to the bottom of the screen.

FIGURE 4-7   If you haven't yet set up iCloud on your iPhone (left), enter your Apple ID and password or create a new Apple ID. After you set up iCloud, you see the available services (right).

6. Tap the Find My iPhone switch and move it to the On position. Your iPhone displays a dialog box (shown here) confirming that you want the iPhone to be tracked.

7. Tap the OK button to close the confirmation dialog box.

8. Tap the Settings button to return to the Settings screen.

> This enables Find My iPhone features, including the ability to show the location of this iPhone on a map.
>
> **Allow**
>
> **Cancel**

Locate Your iPhone with Find My iPhone

After turning on Find My iPhone, you can locate your iPhone at any time from any computer or device that has an Internet connection. To locate your iPhone with Find My iPhone, follow these steps:

1. Open your web browser—for example, Internet Explorer on Windows, Safari on a Mac (or on Windows), or Firefox on most any operating system.

2. Go to www.icloud.com.

3. Log in using your Apple ID. The iCloud home screen appears (see Figure 4-8).

FIGURE 4-8 On the iCloud home screen, click the Find My iPhone icon.

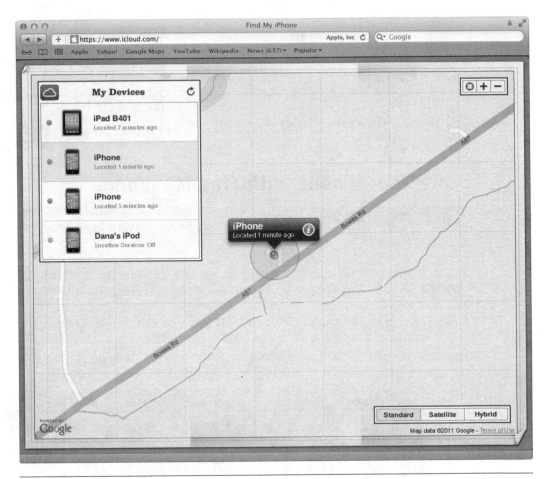

FIGURE 4-9 On the Find My iPhone screen, go to the My Devices list box and click the iPhone or other iOS device you want to locate.

4. Click the Find My iPhone icon to display the Find My iPhone screen (see Figure 4-9).
5. In the My Devices list box, click the iPhone or other iOS device you want to locate. As long as Find My iPhone is able to track your iPhone, its location appears on a map of the area it's in.

 If your iPhone is your only iOS device, it will already be selected in the My Devices list box.

6. If necessary, change the map display so that you can see the location better:
   - Click the Center button (the circular icon that looks like a telescopic sight) in the upper-right corner of the Find My iPhone window to center the map on the iPhone's location.

- Click the + button to zoom in, or click the – button to zoom out.
- Click the Standard button to display a standard map like the one shown in Figure 4-9.
- Click the Satellite button to display a map of satellite imagery.
- Click the Hybrid button to display the standard map's names on the satellite map.

7. Click the *i* button in the iPhone's location button to display the Info dialog box (shown here). You can then take actions with your iPhone as described in the next project.

Project 31: Lock or Wipe Your iPhone After It Gets Lost or Stolen

As you saw in the previous section, you can use the Find My iPhone feature to locate your iPhone when it goes missing.

Once you know your iPhone's geographical location, you'll probably have a better idea of what's happened to it and what you should do. For example:

- If you realize you've left your iPhone in your car or your office, you can cancel the APB and go retrieve it.
- If you can see that your iPhone is following your spouse's usual route to work, you can display a message on the iPhone asking him or her to pull a U-turn and give it back.
- If you detect that your iPhone has entered terra incognita, you'll probably want to make sure it's locked, then display a message on it—and then wipe the iPhone's data if you don't get a positive response.

In the following sections, we'll look at your options in turn.

Lock Your iPhone with a Passcode

What you'll often want to do first is lock your iPhone if either you hadn't applied a passcode lock before or you suspect that the iPhone may have been unlocked when whoever has it picked it up.

To lock your iPhone with a passcode, follow these steps on the Find My iPhone screen:

1. Click the *i* button in the iPhone's location button to display the Info dialog box.
2. Click the Remote Lock button to display the Remote Lock dialog box (shown here).

3. Click the buttons for the four-digit passcode you want to apply.
4. Click the Lock button to apply the passcode to the iPhone.

The passcode takes effect almost immediately—just as soon as it passes through the Internet and over the air to your iPhone.

Display a Message on Your iPhone

The next capability the Find My iPhone feature offers is to display a message on your iPhone's screen. You can also choose whether to play a sound on the iPhone to cause whoever has the iPhone, or is near it, to look at its screen.

To display a message on your iPhone, follow these steps on the Find My iPhone screen:

1. Click the *i* button in the iPhone's location button to display the Info dialog box.
2. Click the Play Sound Or Send Message button to display the Send Message dialog box (shown here).

3. Type your message in the Message box. For example, type a message requesting the finder to call your other phone number to arrange the iPhone's return.
4. If you want to play a sound, make sure the Play Sound switch is set to the On position.
5. Click the Send button to send the message and to play the sound (if you chose to do so).

Wipe Your iPhone's Contents Remotely

If you've exhausted your other options for recovering your iPhone, you can wipe the data it contains to make sure nobody else can read it.

 Treat wiping your iPhone's contents as a last resort, because wiping them means that you can no longer locate the iPhone. Unless you're wiping the iPhone for practice, or you have exceptional luck, you'll never see your iPhone again.

To wipe your iPhone's contents, follow these steps:

1. Click the *i* button in the iPhone's location button to display the Info dialog box.
2. Click the Remote Wipe button to display the Wipe iPhone dialog box (shown here).

3. Click the Wipe iPhone button, and wave a fond farewell to your iPhone.

 To wipe its data, your iPhone simply deletes the key used to encrypt the data. The key is tiny, so this deletion takes only the blink of an eye and renders the data unreadable, even though it is still on the iPhone. Early iPhones used to perform wiping by writing rubbish data over the real data, but this took a while—up to several hours for a high-capacity iPhone—and could run the battery out of power before the wipe was complete.

5 Cellular, Wi-Fi, and Remote Geekery

So far in this book, we've geeked out on music, photography, work, and security. Now it's time to turn our attention to your iPhone's cellular capability, Wi-Fi network connections, and ability to control your computer remotely.

Most of the iPhones in the whole wide world come locked to a particular carrier—for example, AT&T or Verizon in the United States or Rogers in Canada. We'll start by looking at how you can unlock your iPhone from your carrier so that you can connect it to a different carrier's network instead.

After that, I'll show you how to share your iPhone's Internet connection with your computers and devices, so that you can get them online no matter where you are, and how to take control of your PC or Mac from your iPhone.

Finally, you'll learn how to connect your iPhone to your company network across the Internet using a virtual private network (VPN), and how to make Voice over IP calls instead of cellular calls.

Project 32: Unlock Your iPhone from Your Carrier

If you bought your iPhone from a particular carrier, the iPhone will be locked to that carrier's network. So you can't just eject your current SIM card (the subscriber identity module, the card that gives your iPhone its cellular identity), pop in a new SIM card for another carrier, and start using that carrier's network. Instead, you need to unlock the iPhone so that you can use it freely.

How you unlock your iPhone depends on which country you're in, which carrier your iPhone is currently locked to, and which type of contract you're on. Because unlocking has many variables, this project explains the essentials of unlocking your iPhone but leaves the details up to you.

 Before deciding to unlock your iPhone, make sure you understand how locking works and what the consequences of unlocking may be.

Understand Why Most iPhones Are Locked

Typically, a carrier offers only locked iPhones to make sure you stay with that carrier for the duration of your contract (and perhaps longer). As you know, most carriers sell the iPhone at a hefty discount from its headline cost, and then charge you for a monthly plan for a year or two. By the end of the contract, the carrier is ahead on the hardware cost.

If you want to avoid a long contract, you can buy an unlocked iPhone, install in it a SIM card for your preferred GSM carrier's network, and use it for as long as you like. The unlocked iPhone is much more expensive upfront than the locked iPhone—for example, at this writing an unlocked 64GB iPhone 4S costs $849 in the United States, whereas you'd pay $399 for a locked iPhone from AT&T. But you can save in the long term by paying only for the usage you need rather than paying a fixed fee every month on a contract. And you can resell the iPhone at any time without being tied by a contract.

Apple sells unlocked iPhones only in some countries; at this writing, the countries include the United States, Canada, and the United Kingdom. But because the unlocked iPhone 4S is a GSM phone, you can use it in any country that has a GSM carrier. So if you can't buy an unlocked iPhone in your country, you can buy it from a U.S., UK, or Canadian vendor. Check beforehand to see if you'll need to pay import duties, and factor them into your calculations.

Understand How iPhone Locking Works

So that's *why* carriers lock the iPhones (and other phones)—but how does the locking work?

The locking is called *SIM locking*, because it uses the SIM card. A carrier can lock an iPhone to accept only SIM cards that have an approved International Mobile Subscriber Identity (IMSI). For example, the carrier can lock the iPhone so that it'll work only if the SIM has the carrier's own network code. Or the carrier can use the Mobile Station Identification Number (MSIN—the SIM number) to lock the iPhone so it will work only with a particular SIM card.

DOUBLE GEEKERY

Unlocked GSM Phones Won't Work on Verizon and Sprint

GSM and CDMA are the terms for the two most widely used technologies for cellular phones. GSM stands for Global System for Mobile Communications, and CDMA stands for Code Division Multiple Access.

At this writing, Verizon and Sprint use CDMA rather than GSM. Because of this, GSM phones (such as the unlocked iPhone 4S and iPhone 4 models) won't work on Verizon and Sprint.

Apple builds SIM locking into the iPhone because Apple wants the carriers to support the iPhone and sell it vigorously.

Understand the Ways of Unlocking Your iPhone

There are four main ways of unlocking your iPhone:

- Get your carrier to unlock it for you over the air.
- Get the master code from the carrier, and then unlock the iPhone yourself.

DOUBLE GEEKERY

Find Your iPhone's IMEI

To get your iPhone unlocked, you may need to know the iPhone's International Mobile Equipment Identity (IMEI). This is a 15-digit decimal number that's unique to your iPhone.

You can find your iPhone's IMEI either on the iPhone itself or in iTunes with the iPhone connected.

On the iPhone itself, you have two options for finding the IMEI. The first option is to press the Home button, choose Settings | General | About, scroll down to the bottom of the second box, and then look at the IMEI readout. The second option is to open the Phone app, tap the Keypad button, dial *#06#, and tap the Call button. The IMEI comes up in the middle of the screen.

In iTunes, follow these steps:

1. Click the iPhone's entry in the Devices category in the Source list to display the iPhone's control screens.
2. If the Summary screen isn't displayed, click the Summary tab to display it.
3. In the top box, click the Phone Number readout. iTunes displays the IMEI readout instead, as shown here.

iPhone

Name: gPhone
Capacity: 57.42 GB
Software Version: 5.0.1
Serial Number: C39GJ4G4CTDM
IMEI: 012940201417776

4. Click the IMEI readout to display the ICCID readout. This is the Integrated Circuit Card ID, a number of up to 19 digits that works as the primary account number to identify the SIM card. Click a third time to display the phone number again.

The master code for unlocking an iPhone is also sometimes called the *network code key* or the *multilock code*.

- Run software on your computer, connect the iPhone, jailbreak it, and then unlock it.
- Connect your iPhone to a hardware unlocking device, and unlock it.
  We'll look at each of these possibilities in turn.

Get Your Carrier to Unlock Your iPhone

As discussed earlier in this section, the easiest option for getting an unlocked iPhone is simply to buy an iPhone that isn't locked. This approach has a heavy upfront expense, and if you already have an iPhone, you probably won't want to consider it—at least, not until Apple releases the next iPhone.

The next option is to get your carrier to unlock your iPhone for you. If this option is open to you, take it—it's far preferable to messing around with a software unlock or hacking the SIM card.

Some carriers (such as AT&T) will not unlock iPhones. Most carriers who do unlock iPhones will unlock an iPhone only at the end of its contract or on payment of a hefty fee. Getting your carrier to unlock the iPhone is fully legal and aboveboard in all countries.

If your carrier does unlock iPhones, you may have to wait until the end of the contract, you may have to pay a fee, or both. If you have to wait until the end of the contract, you may well have upgraded already to the next iPhone.

Some carriers unlock the iPhone over the air, which can take a day or two to implement. Other carriers give you an unlock code that you enter using the iPhone's keypad.

In some countries, you can also find services online that unlock iPhones (and other phones) for you. These services work by submitting your iPhone's IMEI to Apple, just as the carrier would, and requesting an unlock. The cost varies depending on the iPhone, the carrier, and the contract. Most services that are worth using aren't cheap, but they're effective. The unlocking procedure takes several days to complete.

Use Software to Unlock Your iPhone

If your carrier won't unlock your iPhone, you need to take matters into your own hands. This means unlocking the iPhone by jailbreaking it (as described in Chapter 6) and then using an unlocking application such as Ultrasn0w.

Apple frequently changes the security arrangements in iOS to prevent unlocking, and the developers of the unlocking software then have to develop new versions—so the exact moves you need to perform to unlock your iPhone vary. But here are the general steps to follow:

1. Find the latest instructions for unlocking by visiting a site such as Redmond Pie (www.redmondpie.com) or by searching online. Make sure the instructions are for your iPhone model rather than other iPhones.
2. Download an unlocking tool such as Sn0wbreeze from a site such as www .idownloadblog.com/iphone-downloads/.
3. Jailbreak your iPhone following the instructions in Project 38 in Chapter 6.
4. Follow the instructions for the unlocking tool to unlock your iPhone.

 Unlocking your iPhone via software or hardware is legal in the United States and the United Kingdom but illegal in some countries. If in doubt whether it's legal in your country, check online.

Use a Hardware Unlocking Device

Another way of unlocking an iPhone is to use a hardware unlocking device. These devices are typically operated by companies that unlock phones as part of their business rather than something you'd buy yourself to unlock a single iPhone. You take your iPhone to such a service, pay (incvitably), and have the company unlock it for you.

 You can also get unlocking SIM cards for unlocking the iPhone. Some work, others don't, so look for good reviews before buying. These SIM cards are specific not only to the iPhone model but also to the iPhone's baseband version—so make sure you get exactly the card you need. To find the baseband version, choose Settings | General | About, and then scroll down and look at the Modem Firmware number.

Project 33: Share Your iPhone's Internet Connection with Your Computers and Devices

Your iPhone can not only get a high-speed Internet connection through the cellular network, but it can also share that connection with your computer or other devices. This capability is great for when you're on the road and need to get your computer online where no Wi-Fi connection is available. But you can also use it for home Internet access if your data plan is generous enough.

Sharing the iPhone's Internet connection used to be called *Internet tethering*, and some people still use that term. In iOS 5, the feature for sharing the Internet connection is called Personal Hotspot. You can connect up to five computers or other devices at a time using Personal Hotspot. You can connect a single computer via USB or connect multiple computers and devices via Wi-Fi or Bluetooth. In this section, we'll look at how to use USB and Wi-Fi, which are the two most useful connections.

USB gives the fastest connection to Personal Hotspot—but it works for only one computer at a time. Wi-Fi gives good speeds and is the best choice for connecting multiple devices. Bluetooth gives slower speeds and requires pairing your iPhone with the computer or device, so it is best used only when you have no other means of connection.

Set Up Personal Hotspot

To set up Personal Hotspot on your iPhone, follow these steps:

1. Press the Home button to display the Home screen.
2. Tap the Settings icon to display the Settings screen.
3. Scroll down to the third box, and then tap the General button to display the General screen.
4. Tap the Network button to display the Network screen.
5. Tap the Personal Hotspot button to display the Personal Hotspot screen (shown on the left in Figure 5-1).
6. Tap the Personal Hotspot switch and move it to the On position. The Personal Hotspot screen shows that the network is discoverable under the name you've given your iPhone.

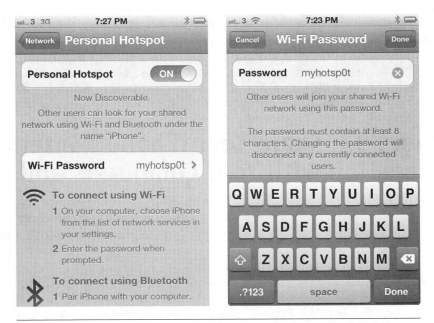

FIGURE 5-1 On the Personal Hotspot screen (left), move the Personal Hotspot switch to the On position, and then tap the Wi-Fi Password button to display the Wi-Fi Password screen (right). Type in the password you want to use, and then tap the Done button.

7. Look at the default password on the right side of the Wi-Fi Password button. If you want to change it, tap the Wi-Fi Password button, and then type the new password on the Wi-Fi Password screen (shown on the right in Figure 5-1). Tap the Done button to return to the Personal Hotspot screen.
8. Tap the Network button to return to the Network screen.
9. Tap the General button to return to the General screen.
10. Tap the Settings button to return to the main Settings screen. You'll see that Personal Hotspot now appears on it under the Wi-Fi item, giving you quick access to the settings for turning Personal Hotspot on and off.

Now that you've turned on Personal Hotspot, you can connect your computers or devices to it.

Connect a Computer or Device to Personal Hotspot via Wi-Fi

To connect a computer or device to Personal Hotspot via Wi-Fi, you need only connect via Wi-Fi to the Personal Hotspot wireless network, just as you would connect to any other wireless network.

The Personal Hotspot wireless network has your iPhone's name and uses the password that appears on the Personal Hotspot screen.

Connect a Single Computer to Personal Hotspot via USB

Instead of connecting via Wi-Fi, you can connect a single computer to Personal Hotspot by using your iPhone's USB cable.

Connect a Windows PC to Personal Hotspot via USB

When you connect your iPhone via USB to a Windows PC, and Personal Hotspot is enabled on the iPhone, Windows automatically detects the iPhone's Internet connection as a new network connection. The first time this happens, Windows automatically installs the driver for the connection and displays the Driver Software Installation dialog box to let you know it has done so. Click the Close button to close the dialog box.

Next, Windows displays the Set Network Location dialog box (see Figure 5-2), asking you whether this new network is a Home Network, a Work Network, or a Public Network. Normally, you'll want to click the Home Network button here.

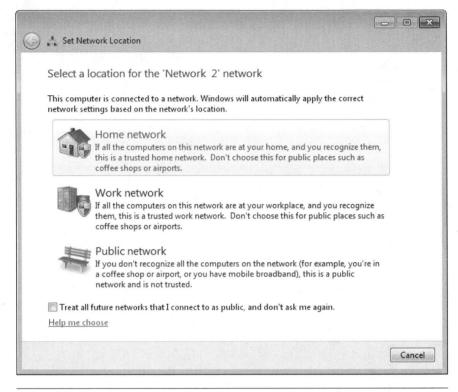

FIGURE 5-2 In the first Set Network Location dialog box, click the Home Network button to tell Windows that the Personal Hotspot network is safe to use.

Windows then sets up the network. When it has done so, it displays another Set Network Location dialog box (see Figure 5-3) confirming the network location.

Click the Close button to close the Set Network Location dialog box. The connection is now ready for you to use.

 An easy way to check that the Internet connection is working is to open Internet Explorer and see if it can load your home page.

Connect a Mac to Personal Hotspot via USB

When you connect your iPhone via USB to a Mac, and Personal Hotspot is enabled on the iPhone, the Mac automatically detects the iPhone's Internet connection as a new network connection. The first time this happens, Mac OS X automatically displays the

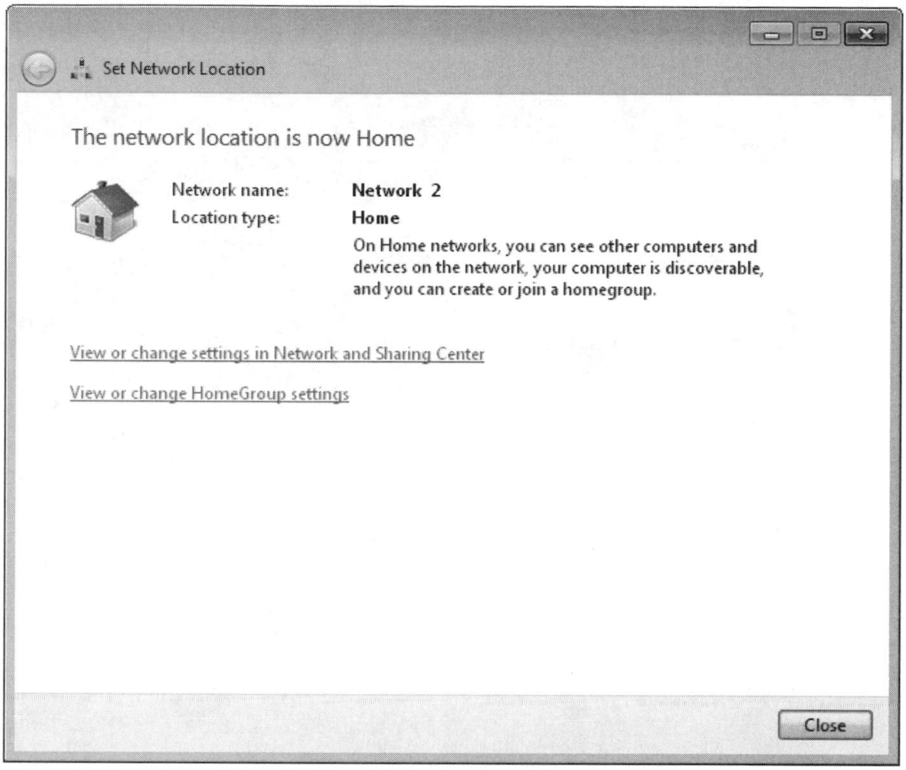

FIGURE 5-3   In the second Set Network Location dialog box, click the Close button. You can then start using the network.

Network preferences pane in System Preferences (see Figure 5-4) so that you can set up the network.

Click the iPhone USB interface in the left box, and then click the Apply button. Mac OS X assigns an IP address to the iPhone USB interface, and then displays the details (see Figure 5-5).

Press ⌘-Q or choose System Preferences | Quit System Preferences to quit System Preferences. You can now start using the Internet connection.

If you want to check that the Internet connection is working, open Safari and see if your home page appears.

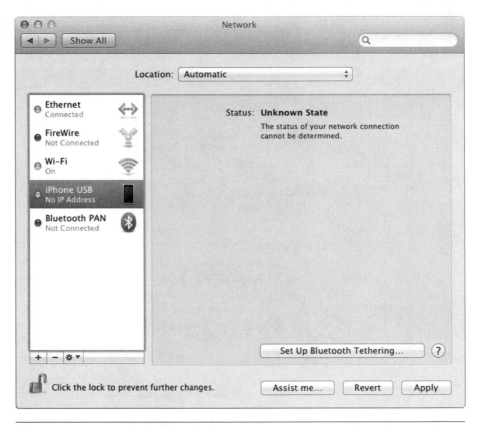

FIGURE 5-4 In the Network preferences pane in System Preferences, click the iPhone USB interface in the left box, and then click the Apply button.

Turn Off Personal Hotspot

When Personal Hotspot is on with no computers or devices connected to it, the only way to tell it's on is that the Personal Hotspot switch on the Personal Hotspot screen is in the On position.

When any computers or devices are connected to Personal Hotspot, your iPhone displays a blue bar across the top of the screen, as shown here.

To turn off Personal Hotspot, follow these steps:

1. Press the Home button to display the Home screen.
2. Tap the Settings icon to display the Settings screen.

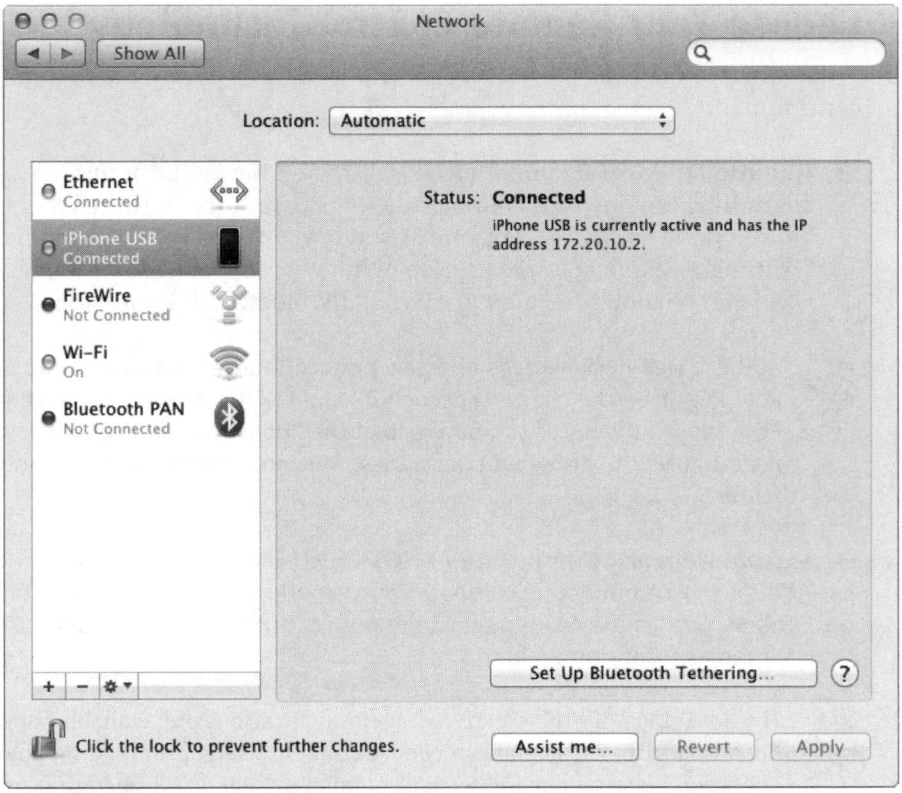

FIGURE 5-5   Mac OS X assigns an IP address to the iPhone USB interface to enable your Mac to use the iPhone as a network connection.

3. Tap the Personal Hotspot button to display the Personal Hotspot screen.
4. Tap the Personal Hotspot switch and move it to the Off position.

Project 34: Control Your PC or Mac from Your iPhone

If you use your iPhone to get work done no matter where you happen to be, you'll definitely want to make the most of your iPhone's capability to control computers remotely. In this project, I'll show you how to reach out from your iPhone and take control of a PC or Mac anywhere on the Internet.

First, we'll get you the remote-control software you need for your iPhone. Then we'll set up your PC or Mac for remote control. After that, you'll be ready to take control of your PC or Mac from your iPhone across the Internet.

Choose Your Remote Control Technology

There are two main technologies for connecting to a computer remotely and controlling it:

- **Remote Desktop Protocol (RDP)**   RDP is Microsoft's proprietary protocol for controlling Windows PCs remotely. RDP is part of the Terminal Services feature built into the "business" versions of Windows: Windows 7 Professional, Windows 7 Ultimate, Windows 7 Enterprise, Windows Vista Business, Windows Vista Ultimate, Windows Vista Enterprise, and Windows XP Professional.

 RDP is a well-designed and effective protocol that enables you to work remotely on your PC. Given the choice between RDP and VNC for connecting to your Windows PC, choose RDP. But if you have one of the "Home" versions of Windows, you will need to use VNC instead, because these versions don't have the Remote Desktop feature.

- **Virtual Network Computing (VNC)**   VNC is a protocol originally developed by AT&T for controlling one computer from another computer. VNC is built into Mac OS X as part of the Screen Sharing feature, but you can add a VNC server to a Windows PC if you need to.

 The advantage of VNC is that VNC client applications are available for all major operating systems, so you can connect to a VNC server running on any major operating system from a VNC client running on any major operating system.

You can find plenty of RDP client apps and VNC client apps on the App Store. In this project, we'll use the apps Mocha RDP and Mocha VNC. Each works well, is comparatively inexpensive at $5.99, and has a free Lite version (supported by ads) that you can try out to see if you want to pay for the full version.

Set Up Your PC for Remote Control

To set up your PC for remote control, follow these steps:

1. Press WINDOWS KEY–BREAK to display the System window. You can also click the Start button, right-click the Computer item to display the context menu, and then click the Properties item on it.
2. In the left column, click the Remote Settings link to display the Remote tab of the System Properties dialog box (see Figure 5-6).
3. In the Remote Desktop box, select the Allow Connections From Computers Running Any Version Of Remote Desktop (Less Secure) option button.

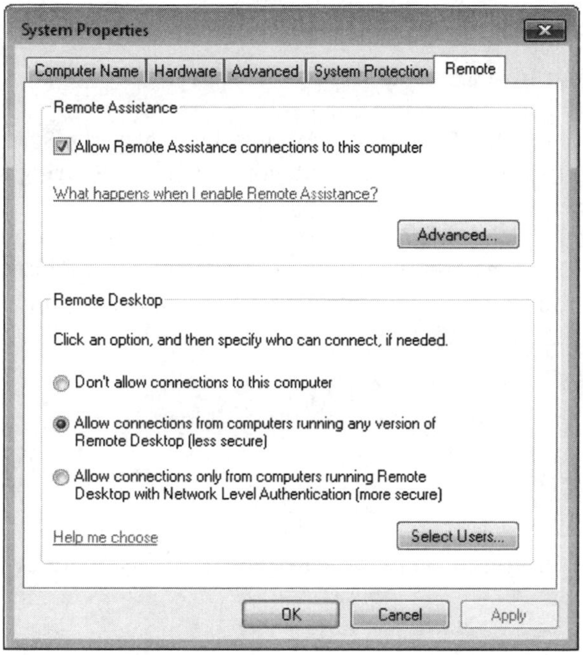

System Properties

| Computer Name | Hardware | Advanced | System Protection | Remote |

Remote Assistance

☑ Allow Remote Assistance connections to this computer

What happens when I enable Remote Assistance?

Advanced...

Remote Desktop

Click an option, and then specify who can connect, if needed.

○ Don't allow connections to this computer

◉ Allow connections from computers running any version of Remote Desktop (less secure)

○ Allow connections only from computers running Remote Desktop with Network Level Authentication (more secure)

Help me choose   Select Users...

OK   Cancel   Apply

FIGURE 5-6   On the Remote tab of the System Properties dialog box, select the Allow Connections From Computers Running Any Version Of Remote Desktop (Less Secure) option button.

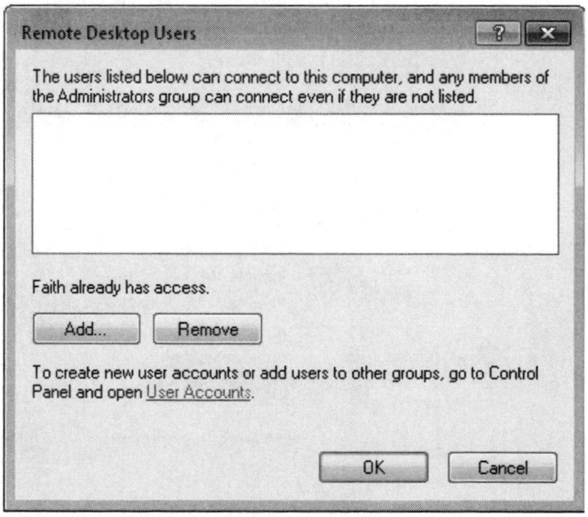

4. Click the Select Users button to display the Remote Desktop Users dialog box (shown here).

5. Verify that your name appears above the Add button with the message "already has access." If not, click the Add button and use the Select Users dialog box to add yourself to the list of users who can connect via Remote Desktop.

6. Click the OK button to close the Remote Desktop Users dialog box.

7. Click the OK button to close the System Properties dialog box.

8. Click the Close button (the × button) to close the System window.

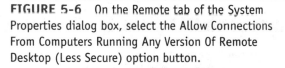

Remote Desktop Users

The users listed below can connect to this computer, and any members of the Administrators group can connect even if they are not listed.

Faith already has access.

Add...   Remove

To create new user accounts or add users to other groups, go to Control Panel and open User Accounts.

OK   Cancel

Set Up Your Mac for Remote Control

To set up your Mac for remote control, follow these steps:

1. Choose Apple | System Preferences to display the System Preferences window.
2. In the Internet & Wireless category, click the Sharing icon to display the Sharing preferences pane.
3. In the left pane, click the Screen Sharing item (but don't select its check box yet) to display the Screen Sharing options (shown in Figure 5-7).
4. Click the Computer Settings button to display the dialog box shown here.

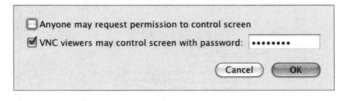

5. Make sure the Anyone May Request Permission To Control Screen check box is cleared.

FIGURE 5-7   Click the Screen Sharing item in the left pane of the Sharing preferences pane to display the controls for setting up sharing.

6. Select the VNC Viewers May Control Screen With Password check box.
7. In the text box, type the password you will use from VNC.
8. Click the OK button to close the dialog box.
9. In the Allow Access For area, select the All Users option button or the Only These Users option button, as appropriate. Normally, you will want to select the Only These Users option button, and then either leave the Administrators group in the list box (where it appears by default) or click the Add (+) button and add yourself as the user who is allowed to access the Mac via Screen Sharing.
10. Now that you have specified who may connect, select the Screen Sharing check box in the left pane.
11. Press ⌘-Q or choose System Preferences | Quit System Preferences to quit System Preferences.

Take Control of Your PC with Your iPhone

Now that you've set up your PC to accept RDP connections, you can connect to it from your iPhone using the Mocha RDP app. First, you launch the RDP app and set up the details of the connection. Then you establish the connection and get to work. And when you finish using the connection, you disconnect from the computer or log off Windows.

Launch the Mocha RDP App and Create a Connection

To create a connection, follow these steps:

1. Launch the RDP app from your iPhone's Home screen as usual. The app then displays the Mocha RDP screen (shown on the left in Figure 5-8) with the Configure button selected as a hint you should tap it.

 The RDP app offers many settings that you can use to adjust how the app behaves. In this section, we'll set only the essential settings, such as the computer's address and the screen resolution. When you have time, explore the other options and see which suit you.

2. Tap the Configure button to display the first Configure screen (shown on the right in Figure 5-8).
3. Tap the New button to start creating a new configuration file. The RDP app displays the second Configure screen (shown on the left in Figure 5-9).
4. Tap the > button at the right end of the PC Address button to display the Lookup screen (shown on the right in Figure 5-9).

 If you know your PC's computer name or IP address, tap in the <required> placeholder on the PC Address button to place the insertion point and bring up the onscreen keyboard. You can then type the computer name or IP address.

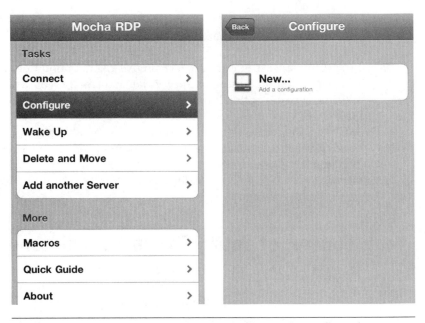

FIGURE 5-8 On the Mocha RDP screen (left), tap the Configure button to display the first Configure screen (right). Then tap the New button to start setting up a new connection.

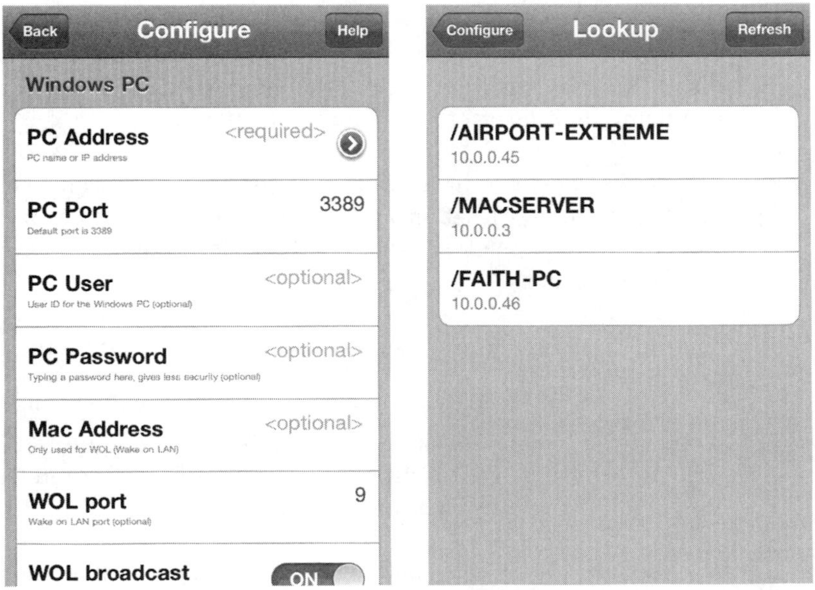

FIGURE 5-9 On the second Configure screen (left), tap the PC Address button to display the Lookup screen (right), and then tap the name of the computer you want to connect to.

5. Tap the name of the computer you want to connect to. The RDP app returns you to the second Configure screen, where the PC Address button now shows the computer's name.

6. If your PC is using a nonstandard port, tap the PC Port button, and then type the port number.

7. If you want the RDP app to store your username, tap the PC User button, and then type your username.

8. Similarly, if you want the RDP app to store your password, tap the PC Password button, and then type your password.

9. Scroll down to the second box (shown on the left in Figure 5-10), and then tap the Screen Size button. On the PC Screen Size screen (shown on the right in Figure 5-10) that appears, tap the button for the resolution you want. You can tap the > button at the bottom to set a custom resolution—for example, 960 × 640 pixels to fill the iPhone's screen.

10. When you finish choosing settings for the connection, tap the Back button to return to the first Configure screen. The connection appears as a button, as shown in the illustration.

11. Tap the Back button to return to the Mocha RDP screen.

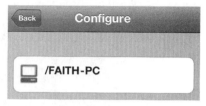

FIGURE 5-10   From the lower part of the second Configure screen, tap the Screen Size button to display the PC Screen Size screen (right), and then tap the button for the resolution you want.

Connect to Your PC

From the Mocha RDP screen, follow these steps to connect to your PC:

1. Tap the Connect button. The RDP app connects to your PC.

 If you've set up only one connection in the RDP app, the app automatically opens that connection when you tap the Connect button on the Mocha RDP screen. If you've set up multiple connections, the RDP app displays the Connect To screen. Tap the button for the computer you want to connect to.

2. If you didn't enter your username and password, the RDP app displays the Windows login screen. Tap your username to display the Password field, as shown here, and then tap the keyboard icon to display the keyboard, as shown next. Type your password, and then tap the Return button to enter it.

3. The RDP app then displays your Windows Desktop (see Figure 5-11), and you can start working on it. These are the main moves you'll need:
   - **Click**   Tap with your finger.
   - **Double-click**   Double-tap.

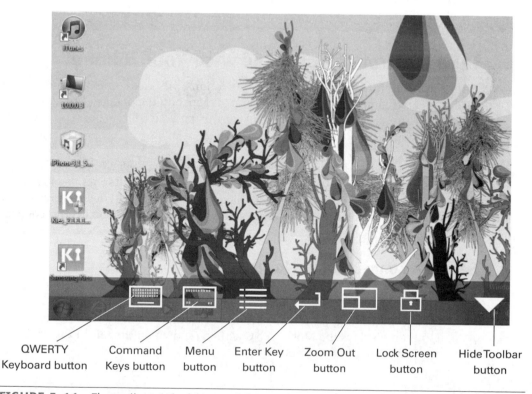

QWERTY   Command   Menu   Enter Key   Zoom Out   Lock Screen   Hide Toolbar
Keyboard button Keys button button   button   button   button   button

FIGURE 5-11   The toolbar at the bottom of the screen in the RDP app gives you quick access to the keyboard, the menu, and the command for zooming out.

- **Right-click**   Tap and hold for a second.
- **Zoom in**   Place your thumb and finger (or two fingers) together on the screen, and then pinch outward.
- **Zoom out**   Place your thumb and finger (or two fingers) apart on the screen, and then pinch them together.
- **Scroll**   Tap and drag your finger to move the displayed part of the screen in that direction.

Disconnect from or Log Off Your PC

When you finish using your PC, you can either disconnect from it or log off:

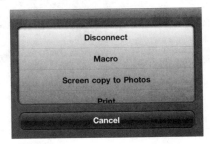

- **Disconnect**   In the RDP app, tap the Menu button to display the Menu screen (shown here), and then tap the Disconnect item. The RDP app disconnects from your PC, but your user session keeps running. So if you connect again, you can pick up your work where you left off.
- **Log off**   In the RDP app, tap the Start button, and then tap the Log Off button. Windows closes your user session, and the RDP app closes the connection to your PC.

Take Control of Your Mac with Your iPhone

After setting your Mac to accept VNC connections, you can connect to it by using the VNC app. First, you'll launch the VNC app and specify the details of the connection. Then you can make the connection and start using your Mac. When you finish using your Mac, you can disconnect from your Mac.

 VNC uses the Mac's current screen resolution—unlike RDP, VNC cannot change the resolution for display on your iPhone. Because of this limitation, you may want to change the resolution your Mac is using if you plan to use VNC extensively. You can change the resolution either while you're at your Mac or remotely after connecting via VNC.

Set Up a Connection in Mocha VNC

To set up a connection in Mocha VNC, follow these steps:

1. Launch the VNC app from your iPhone's Home screen by tapping its icon. The app then displays the Mocha VNC screen (shown on the left in Figure 5-12) with the Configure button selected.

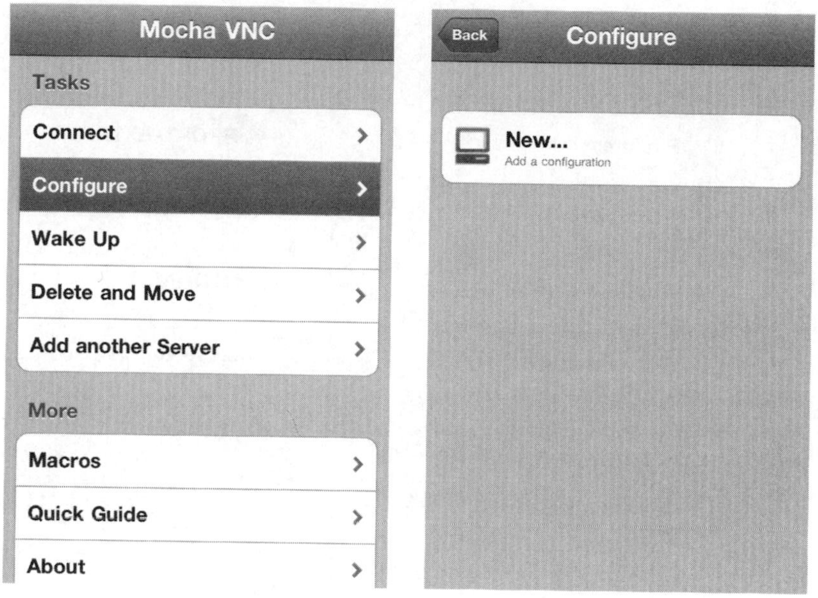

FIGURE 5-12   On the Mocha VNC screen (left), tap the Configure button to display the first Configure screen (right). Then tap the New button to start setting up a new connection.

 The VNC app has many settings for configuring your VNC sessions—for example, choosing which Mac keyboard driver the app uses or controlling whether movements detected by the iPhone's accelerometers scroll the screen in VNC. In this section, we'll set only those settings needed to establish a connection. When you have time, explore the other options and see which you find useful.

2. Tap the Configure button to display the first Configure screen (shown on the right in Figure 5-12).
3. Tap the New button to start creating a new configuration file. The VNC app displays the second Configure screen (shown on the left in Figure 5-13).
4. Tap the > button at the right end of the IP Address button to display the Lookup screen (shown on the right in Figure 5-13).

 If you know your Mac's IP address or computer name, tap in the "write here or (>)" placeholder on the IP Address button to place the insertion point and bring up the onscreen keyboard. You can then type the IP address or computer name.

5. Optionally, tap the VNC Password field and type the password if you want to store it in the connection. If you prefer not to store the password for security reasons, you can provide it when you make the connection.

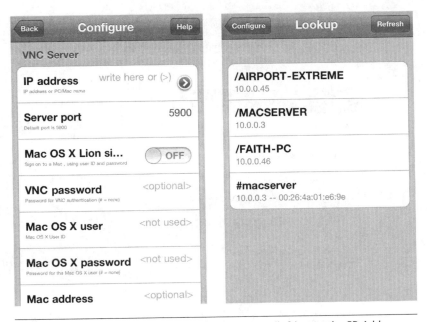

FIGURE 5-13 On the second Configure screen (left), tap the IP Address button to display the Lookup screen (right), and then tap the name of the computer you want to connect to.

 If you're connecting to a Mac that runs Lion (Mac OS X 10.7), you can log in to the Mac remotely instead of merely connecting via Screen Sharing. To do this, tap the Mac OS X Lion Sign On switch and move it to the On position, and then tap the Mac OS X User field and type your username. You can also provide your password by tapping the Mac OS X Password field and typing, or you can wait and provide it when you try to connect to the Mac.

6. Tap the Back button to return to the first Configure screen.
7. Tap the Back button to return to the Mocha VNC screen.

Connect to Your Mac

From the Mocha VNC screen, follow these steps to connect to your Mac:

1. Tap the Connect button:
   - If you have created only one connection, the VNC app connects to it.
   - If you have created multiple connections, the VNC app displays the Connect To screen. Tap the connection you want to use, and the app connects to it.
2. If the VNC app displays the Server Password screen (shown here), type your password, and then tap the OK button.

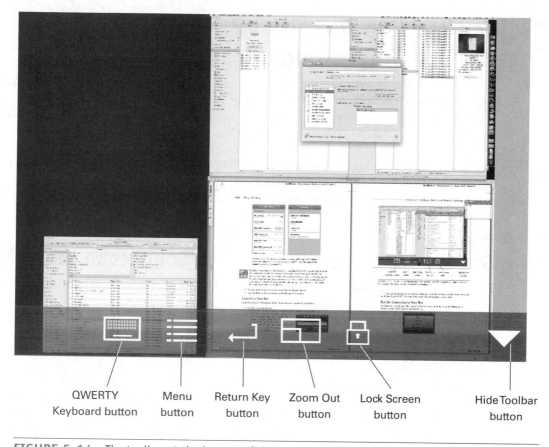

QWERTY
Keyboard button

Menu
button

Return Key
button

Zoom Out
button

Lock Screen
button

Hide Toolbar
button

FIGURE 5-14 The toolbar at the bottom of the screen in the VNC app gives you quick access to the keyboard, the menu, and the command for zooming out.

The app then displays your Mac's desktop, with the toolbar at the bottom overlaid on it (see Figure 5-14). You can then start using the apps on your Mac.

End the Connection to Your Mac

To disconnect from your Mac, tap the Menu button, and then tap the Disconnect button on the Menu screen (shown here).

Project 35: Connect via VPN Across the Internet to Your Company's Network

If you use an iPhone for company business, you may need to connect the iPhone to your company's network so that you can grab your e-mail or exchange data. When you're in the office, you'll probably connect via a wireless network, but when you're out of the office, you can connect across the Internet using a virtual private network, or VPN.

A VPN uses an insecure public network (such as the Internet) to connect securely to a secure private network (such as your company's network). A VPN acts as a secure "pipe" through the insecure Internet, providing a secure connection between your computer (in this case, your iPhone) and your company's VPN server.

Get the Information Needed to Connect to the VPN

To connect to a VPN, you need to know various pieces of configuration information, such as your username, the server's Internet address, and your password or other means of authentication. You also need to know which type of security to use: Layer 2 Tunneling Protocol (L2TP), Point-to-Point Tunneling Protocol (PPTP), or IP Security (IPSec).

Your company's network administrator will provide this information. The administrator may provide it as a written list, which you enter manually in your iPhone, as described a little later in this chapter. But it's easy to get one or more items wrong, so usually an administrator will use the iPhone Configuration Utility (a tool Apple provides for administering the iPhone, iPad, and iPod touch) to create a file called a *configuration profile* that you then install on your iPhone and that does the work for you. We'll start with this easier approach.

 If you're the administrator, you'll find the iPhone Configuration Utility here: www .apple.com/support/iphone/enterprise. There are versions for both Windows and Mac OS X.

Set Up a VPN by Using a Configuration Profile

To set up a VPN on your iPhone by using a configuration profile, all you need to do is get the configuration profile onto your iPhone. Normally, the administrator will either put the configuration profile on your iPhone directly by connecting it to his or her computer via USB or distribute the configuration profile in one of these ways:

- **Via e-mail**   This is an easy way of distributing configuration profiles as long as the administrator knows your e-mail account. But if the configuration profile is for a corporate e-mail account as well as for the VPN, you'll need to use another e-mail account (because the iPhone won't yet be able to access your corporate account).

- **Via a website**   The administrator can place the configuration profile on a website from which you can download it using the iPhone. Typically this will be an internal corporate website or at least a password-protected website, because the configuration profiles aren't encrypted.

Here's how to set up a VPN by installing a configuration profile you've received in an e-mail message or downloaded from a website:

1. Open the configuration profile:
    - If you've received the configuration profile in an e-mail message, as shown in the left screen in Figure 5-15, tap the configuration profile's button. Your iPhone then displays the Install Profile screen, as shown on the right in Figure 5-15.
    - If the configuration profile is posted on a web page, open that page in Safari, and then tap the profile's download link. Your iPhone then displays the Install Profile screen.
2. Look at the information on the Install Profile screen to make sure you want to install the profile. To see more information about the profile, tap the More Details button, which displays the profile's information screen (shown on the left in Figure 5-16). Tap the Install Profile button in the upper-left corner to go back to the Install Profile screen.
3. Check the profile's status: Unsigned, Not Verified, or Verified. See the nearby sidebar "Understand the Unsigned, Not Verified, and Verified Terms on the Install Profile Screen" for an explanation of these terms and advice on how you should treat the profiles they mark.

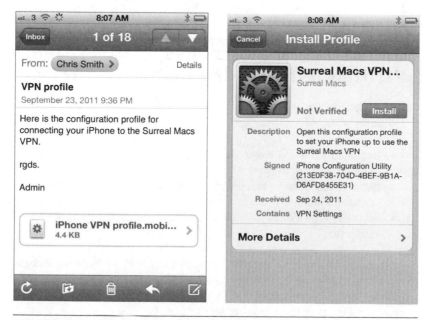

FIGURE 5-15   Tap the configuration profile's button in an e-mail message (left) to display the Install Profile screen (right).

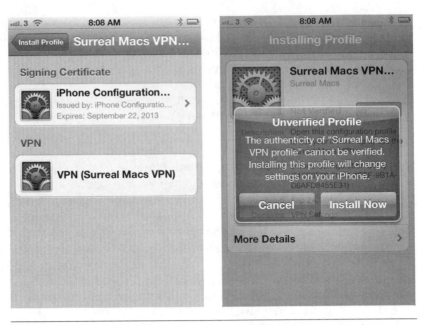

FIGURE 5-16 The profile's information screen (left) shows you the details of what the profile contains—in this case, the signing certificate and the VPN payload. When you tap the Install button on the Install Profile screen to install the profile, your iPhone makes sure you know that installing the profile will change settings on your iPhone (right).

 DOUBLE GEEKERY

Understand the Unsigned, Not Verified, and Verified Terms on the Install Profile Screen

The readout to the left of the Install button on the Install Profile screen shows the profile's status:

- **Unsigned** Whoever created the profile didn't apply a digital signature to the profile to protect it against changes.
- **Not Verified** The creator did apply a digital signature to the profile, but your iPhone can't confirm the digital signature is authentic.
- **Verified** The iPhone has confirmed the digital signature applied to the profile is authentic.

In an ideal world, you'd install only profiles that were verified as coming from whom they claim. But many companies and organizations still use unsigned profiles, so you have a fair chance of running into them. If in doubt, check with an administrator that the profile is safe to install.

4. Tap the Install button on the Install Profile screen to start installing the profile. You'll need to provide your username (as shown on the left in Figure 5-17), password (not shown), and shared secret (as shown on the right in Figure 5-17) to set up the VPN.

5. When the Profile Installed screen appears, tap the Done button. Your iPhone takes you back to where you started the installation—either the e-mail message containing the configuration profile or the web page from which you downloaded the profile.

You can now start using the VPN. Skip ahead to the section "Connect to a VPN," later in this chapter.

Set Up a VPN Manually

If your administrator has supplied you with a list of configuration details for the VPN rather than with a configuration profile, you'll need to set it up the hard way. Because you have to type in all the details on your iPhone, this is somewhat laborious, but you need to do it only once for any connection. Follow these steps:

1. Press the Home button to reach the Home screen.
2. Tap the Settings icon to display the Settings screen.

FIGURE 5-17 Your iPhone walks you through the process of setting up the VPN. You enter first your username on the Enter Username screen (left), next your password on the Enter Password screen (not shown), and then the VPN's shared secret on the Enter Shared Secret screen (right).

3. Scroll down to the third box, and then tap the General button to display the General screen.
4. Tap the Network button to display the Network screen (shown on the left in Figure 5-18).
5. Tap the VPN button to display the VPN screen (shown on the right in Figure 5-18).
6. Tap the Add VPN Configuration button to display the Add Configuration screen, shown on the left in Figure 5-19.
7. Near the top of the screen, click the button for the security type the VPN uses: L2TP, PPTP, or IPSec. The iPhone displays a list of the information required for the connection.
8. Type in the details for the VPN configuration on the screen:

- **Description** This is the name under which the VPN appears in the list of VPNs. Choose a descriptive name that suits you.
- **Server** Type the computer name (for example, macserver.surrealmacs.com) or IP address (for example, 216.248.2.88) of the VPN server.
- **Account** Type your login name for the VPN connection. Depending on your company's network, this may be the same as your regular login name, but in most cases it's different for security reasons.
- **Password** If the administrator has given you a password rather that a certificate (discussed next), you can enter it here and have your iPhone provide it for you each time you connect. For greater security, you can leave the password area blank and enter the password manually each time you connect. This prevents anyone else from connecting using your iPhone,

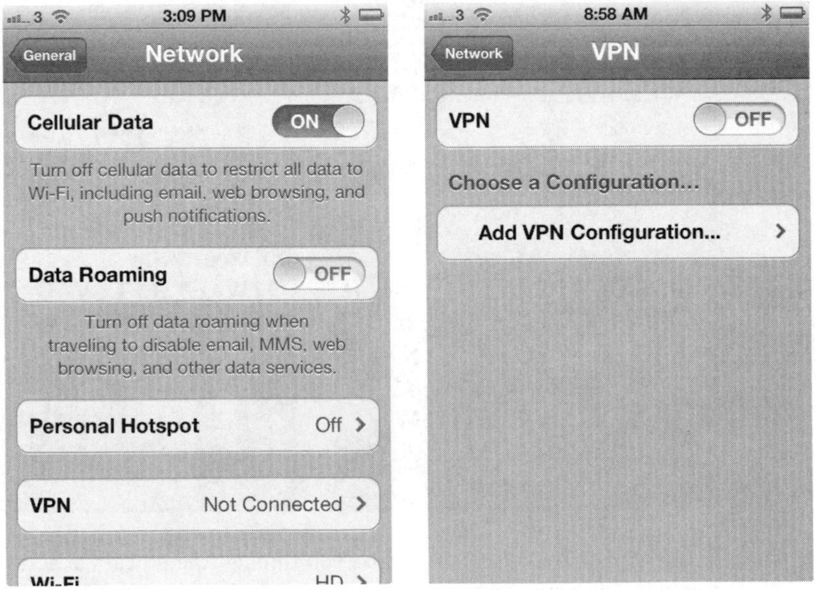

FIGURE 5-18 Tap the VPN button on the Network screen (left) to reach the VPN screen (right).

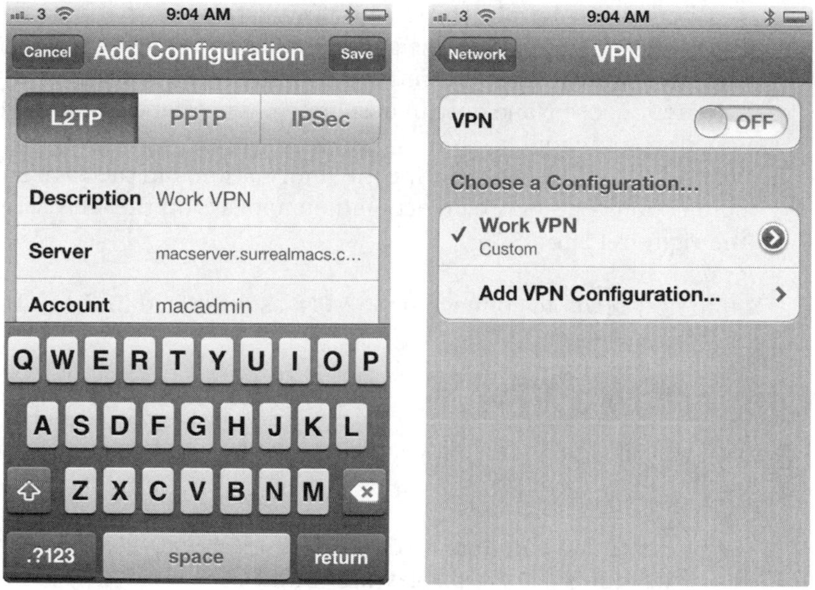

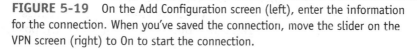

FIGURE 5-19   On the Add Configuration screen (left), enter the information for the connection. When you've saved the connection, move the slider on the VPN screen (right) to On to start the connection.

but it's laborious, especially if your password uses letters, numbers, and symbols (as a strong password should).

- **RSA SecurID**   (PPTP and L2TP only) If the administrator provided you with an RSA SecurID token, move this switch to On to use it. The iPhone then hides the Password field, because you don't need to use a password when you use the token.

- **Use Certificate**   (IPSec only) If the administrator provided you with a configuration profile that installed a certificate for authenticating you on the connection, move this switch to On. To save you from temptation, the switch is available only when a certificate is installed.

- **Secret**   (L2TP only) Type the preshared key, also called the *shared secret*, for the VPN. This preshared key is the same for all users of the VPN (unlike your account name and password, which are unique to you).

- **Group Name**   (IPSec only) Type the name of the group to which you belong for the VPN.

- **Send All Traffic**   (L2TP only) Leave this switch set to On (the default position) unless the administrator has told you to turn it off. When Send All Traffic is on, all your Internet connections go to the VPN server; when it is off, Internet connections to parts of the Internet other than the VPN go directly to those destinations.

- **Encryption Level**   (PPTP only) Leave this set to Auto to have the iPhone try 128-bit encryption (the strongest) first, then weaker 40-bit encryption, and then None. Choose Maximum if you know you must use 128-bit encryption only. Choose None only in desperate circumstances—no sane administrator will recommend it.

9. When you've finished entering the information, tap the Save button to save the connection. The VPN connection then appears on the VPN screen (as shown on the right in Figure 5-19).

You're now ready to connect to the VPN, as described in the next section.

Connect to a VPN

After you've installed or created your VPN connection, you can connect to it quickly and easily. Follow these steps:

1. Press the Home button to reach the Home screen.
2. Tap the Settings icon to display the Settings screen.
3. Start the VPN connection in one of these ways:
   - **If you have only one VPN connection**   On the Settings screen (shown on the left in Figure 5-20), move the VPN switch to the On position.

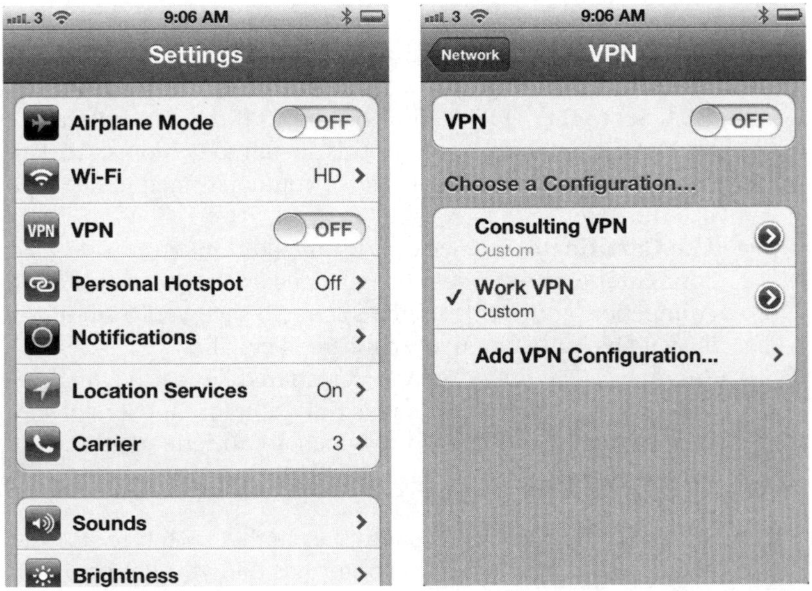

FIGURE 5-20 If you have a single VPN connection, you can turn it on from the Settings screen (left). If you have two or more connections, choose the connection on the VPN screen (right), and then turn it on.

- **If you have two or more VPN connections**   Tap the VPN button to display the VPN screen. In the Choose A Configuration list (shown on the right in Figure 5-20), make sure the correct VPN is selected; if not, tap the one you want, putting a check mark next to it. Then move the VPN switch to the On position.

If the administrator set you up to authenticate yourself with a password, and you chose not to store the password in the VPN connection, you'll be prompted for your password. Enter it, and the iPhone establishes the connection. The Status readout on the VPN screen shows the connection is active (see the left screen in Figure 5-21), and the VPN indicator also appears in the status bar as a reminder you're using the VPN. You can tap the Status readout to see the details of the connection (see the right screen in Figure 5-21), including your iPhone's IP address.

Once you've established the connection, you'll be able to work on the VPN. What exactly you'll be able to do depends on the permissions the administrator has granted you, but you'll typically be able to access your e-mail and shared information resources.

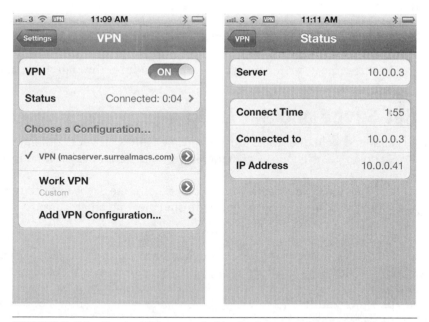

FIGURE 5-21   On the VPN screen (left), the Status readout shows the duration of the connection. You can see further details by tapping the Status button and looking at the Status screen (right). The VPN indicator appears on the status bar as long as the connection is open.

Disconnect from a VPN

When you've finished using the VPN, close any files that you have been using, and then disconnect like this:

1. Press the Home button to reach the Home screen.
2. Tap the Settings icon to display the Settings screen.
3. If you have a single VPN set up, move the VPN switch on the Settings screen to the Off position. Otherwise, tap the VPN button to display the VPN screen, and then move the VPN switch on the VPN screen to the Off position.

Project 36: Make Voice over IP Calls Instead of Cellular Calls

If you find you typically run out of your call minutes on your carrier's plan, or if you bought your iPhone outright and pay for each call, consider making calls using Voice over IP (VoIP) instead of the cellular network. This is easy to do—all you need is to download a suitable VoIP calling app and set up an account.

At this writing, the best app for making VoIP calls is Skype. Skype is a free app, but you pay with Skype Credit for the minutes you use. You can use Skype for plain voice calls, for video calls, and for text messages.

In this project, we'll get Skype, set it up, and run through the essentials of using it.

Get Skype and Set It Up

To get Skype, go to the App Store either using iTunes on your computer or using the App Store app on your iPhone. Search for **Skype**, and then download and install the app.

 If you don't yet have a Skype account, go to the Skype website (www.skype.com) and click the Join Skype button to start creating an account.

Run Skype by tapping the Skype icon on the Home screen. Skype displays a screen prompting you for your Skype name and password, as shown on the left in Figure 5-22. Type in the name and password, and then tap the Sign In button. Skype signs you in and displays the Contacts screen (shown on the right in Figure 5-22).

FIGURE 5-22   On the first Skype screen (left), enter your Skype name and password to start using Skype. On the Contacts screen (right), tap the group containing the contact you want to call.

Make Calls with Skype

After setting up Skype, you can easily place and receive calls:

- **Place a call**   From the Contacts screen, tap one of the groups, and then tap the contact to display its details, as shown on the left in Figure 5-23. You can then place a video call by tapping the Video Call button or a voice call by tapping the Voice Call button.

 If you don't have a contact for the number you want to call, tap the Call tab to display the Call screen, and then dial the number.

- **Receive a call**   When you receive a Skype call, Skype displays the caller's name or number if it's available, as shown on the right in Figure 5-23. Tap the green Accept button (with the picked-up receiver icon) in the lower-right corner of the screen to take the call.

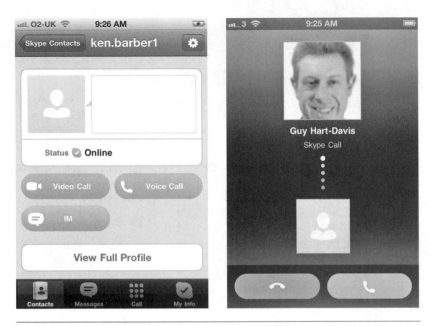

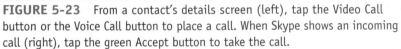

FIGURE 5-23 From a contact's details screen (left), tap the Video Call button or the Voice Call button to place a call. When Skype shows an incoming call (right), tap the green Accept button to take the call.

6 Jailbreaking and Advanced Geekery

So far in this book, we've kept your iPhone within the ecosystem that Apple has created for iOS devices—the iPhone itself, the iPod touch, and the iPad.

This ecosystem is what's known as a "walled garden"—an area that's tightly protected to help you have a safe computing experience in mostly pleasant surroundings. For example, in its normal state, iOS allows you to install only apps from the App Store, which are all approved by Apple. This helps you avoid installing apps that contain malware or that try to ship your credit card details to people who will use them vigorously until you get your bank to issue a cease-and-desist order.

To get outside this walled garden, you need to "jailbreak" your iPhone.

We'll start this chapter by backing up your iPhone's contents so that you can restore them if anything goes wrong during the jailbreak or other moves. Then we'll perform the jailbreak so that we can start performing advanced moves with your iPhone.

Once your iPhone is jailbroken, you'll learn how to find and install unapproved apps and back them up so that you can reinstall them later as needed. You'll explore your iPhone's two partitions and recover any extra space languishing wasted on the OS partition. You'll also apply themes to make your iPhone look different, make Wi-Fi–only apps run over 3G connections when necessary, and play console and arcade games under emulation on your iPhone.

Toward the end of the chapter, we'll get physical. First, we'll open your iPhone so that you can see what's inside it. Then we'll put it back together—with a custom plate on the back to emphasize your uniqueness. And—only if you want—we'll put a near-field communications card in your iPhone so that you can pay for your espresso by waving your iPhone at a payment terminal.

At the very end of the chapter, we'll get brutal: We'll put your iPhone back in its Apple jail.

Let's get started.

Project 37: Back Up Your iPhone's Contents and Settings

Before you jailbreak your iPhone (as described in the next project), back it up to make sure that your precious data and settings are safe. If necessary, you will then be able to restore your data and settings when needed.

DOUBLE GEEKERY

Understand What an iPhone Backup Contains

Before you use iTunes' feature for backing up your iPhone, it's vital you understand what the backup includes and what it doesn't. Otherwise, if you need to restore your iPhone from backup, you may not be able to restore all the files you need.

Your iPhone can contain a huge amount of files—a 64GB iPhone has around 57GB of space available to you—but most of the files will normally be either on your computer or in iCloud as well. For example, if you sync your music, video files, TV shows, and so on with your iPhone, your computer still has these files—so your iPhone backup doesn't need to include them.

So when you back up your iPhone, iTunes syncs your calendars, contacts, notes, text messages, and settings, but not the media files or your iPhone's firmware.

This means that if you create files in third-party apps on your iPhone, you must copy them to your computer or to online storage to keep them safe, because backing up your iPhone doesn't keep copies of them. If you have to erase your iPhone's contents and settings and then restore the iPhone from backup, these files won't be included.

To back up your iPhone, follow these steps:

1. Connect your iPhone to your computer via the USB cable.
2. If iTunes doesn't automatically display the iPhone's control screens, click the iPhone entry in the Devices category in the Source list to display them.
3. If the Summary screen isn't displayed, click the Summary button to display it.
4. In the Backup box, make sure the Back Up To This Computer option button is selected rather than the Back Up To iCloud option button.
5. If you want to encrypt the backup, follow these substeps:
   a. Select the Encrypt iPhone Backup check box. iTunes displays the Set Password dialog box. The illustration shows the Mac version of the Set Password dialog box.

   b. Type a password in the Password box and the Verify Password box.

    c. On the Mac, select the Remember This Password In My Keychain check box if you want Mac OS X to store your password in the Keychain, so that it can enter the password automatically for you.

    d. Click the Set Password button. iTunes starts backing up the iPhone.

6. If you didn't start the backup from the Set Password dialog box, start it by right-clicking (or CTRL-clicking on the Mac) your iPhone's entry in the Source list, and then clicking the Back Up item on the context menu.

Project 38: Jailbreak Your iPhone

After backing up your iPhone as described in the previous project, you're ready to jailbreak it. Jailbreaking your iPhone lets it get out of the walled garden that Apple has penned it in and enables you to install third-party apps and customizations that haven't passed Apple's stringent approvals process.

At this writing, there are several applications you can use to jailbreak your iPhone. Some applications, and some versions of applications, work only with particular models of iPhone, so make sure you choose an application and version that will work with the iPhone model you have.

The Redmond Pie website (www.redmondpie.com) is a good place to find out about jailbreaking tools and techniques. You can also find plenty of other sites by searching using terms such as **jailbreak iPhone 4S**.

DOUBLE GEEKERY

Understand Tethered Jailbreaks and Untethered Jailbreaks

Depending on your iPhone model and the version of iOS it's running, you may be able to choose between a tethered jailbreak and an untethered jailbreak:

- **Tethered jailbreak**   You must connect the iPhone to your computer and use the jailbreaking application each time you want to restart the iPhone in jailbroken mode. We'll perform a tethered jailbreak in this project.
- **Untethered jailbreak**   After you've jailbroken the iPhone, you can restart it without connecting it to your computer. We'll perform an untethered jailbreak in Project 43, which shows you how to save space on your iPhone's OS partition.

As you can see, an untethered jailbreak is far preferable—so you'll probably want one if it's available for your iPhone and version of iOS. But for some iPhone models, versions of iOS, and computer operating systems, you may find that only tethered jailbreaks are available.

In this project, we'll use the application called redsn0w to perform a tethered jailbreak on an iPhone. Follow these steps:

1. Locate and download the appropriate version of redsn0w.
2. Unzip the redsn0w distribution file:
   - **Windows** Click the Start button, and then click your username to open a Windows Explorer window showing your user folder. Double-click the Downloads folder to open it. Right-click the redsn0w zip file, and then click the Extract All item on the context menu to launch the Extract Compressed (Zipped) Folders wizard. Choose the destination folder, make sure the Show Extracted Files When Complete check box is selected, and then click the Extract button.
   - **Mac** Click the Downloads icon on the Dock, and then click the Open In Finder button at the bottom. In the Finder window that opens, double-click the redsn0w zip file to unzip it.
3. Quit iTunes if it's running. You need to do this because if you don't, redsn0w will force iTunes to quit, and the next time you run iTunes, it'll need to check your library to make sure it's okay.
4. Run redsn0w by clicking the redsn0w file in the Windows Explorer window or the Finder window. When redsn0w opens, you'll see the screen shown on the left in Figure 6-1.

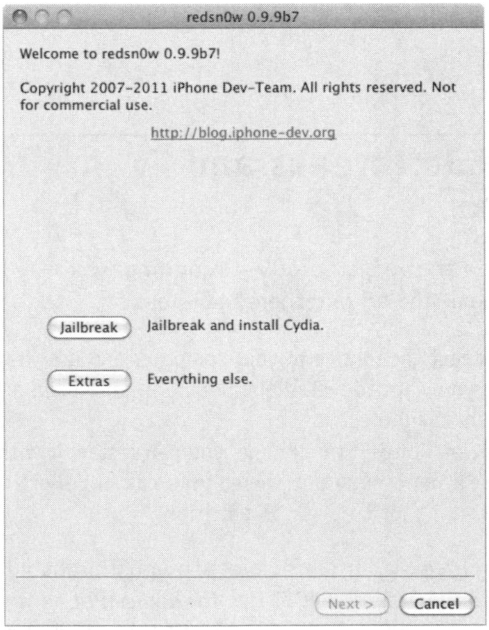

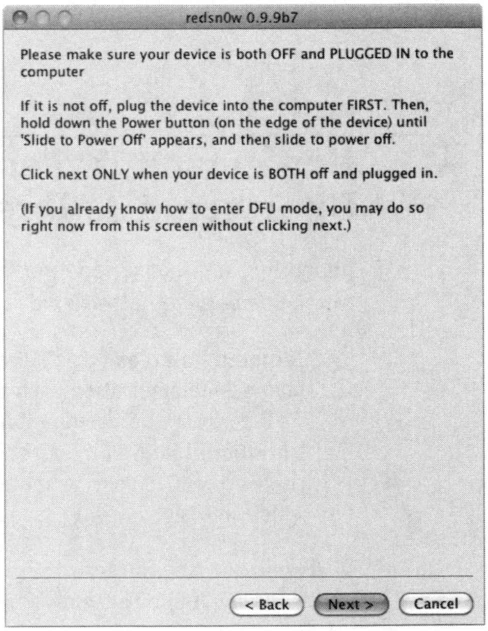

FIGURE 6-1 In the redsn0w window that opens at first (left), click the Jailbreak button. In the window that then opens (right), follow the instructions, and then click the Next button.

5. Click the Jailbreak button. The redsn0w window displays the screen shown on the right in Figure 6-1, telling you to connect your iPhone to your computer and to power it down.

6. Connect your iPhone to your computer.

7. Turn your iPhone off. Hold down the Sleep/Wake button until the Slide To Power Off screen appears, and then slide the slider to the right. Wait until the shutdown progress indicator disappears.

8. Click the Next button to display the screen shown on the left in Figure 6-2, which walks you through putting your iPhone into Device Firmware Upgrade (DFU) mode by pressing and holding the power button, then holding down the Home button as well, and finally continuing to hold down the Home button but releasing the power button. The screen counts down the timing for each step, making the process as simple as possible.

9. After you put your iPhone into DFU mode, redsn0w takes control of your iPhone and analyzes it, as shown on the right in Figure 6-2. redsn0w then displays the Please Select Your Options screen (shown on the left in Figure 6-3).

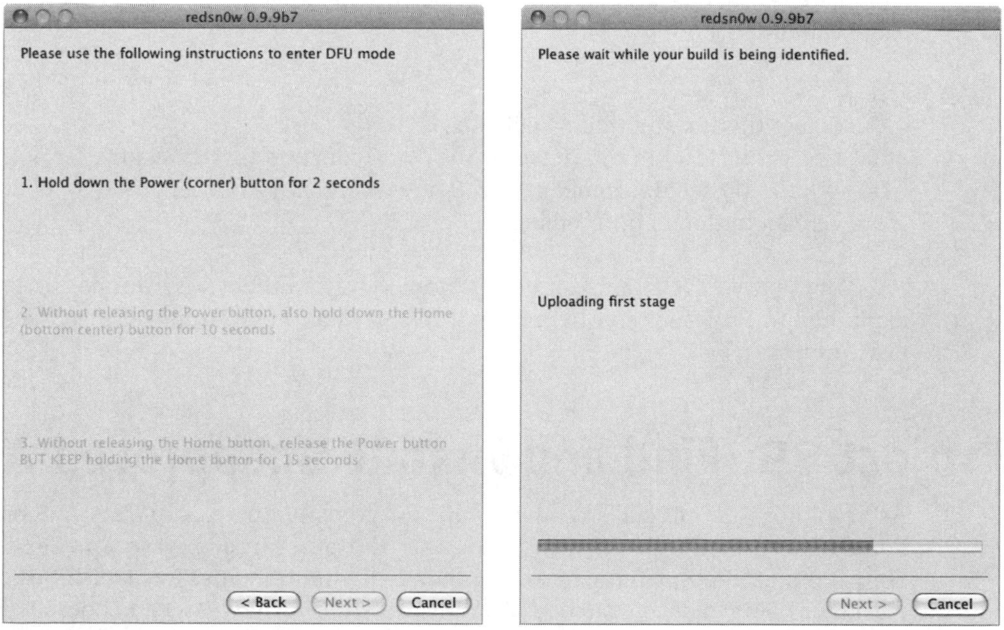

FIGURE 6-2   Follow the instructions and timings on the DFU Mode screen (left) to put your iPhone into Device Firmware Upgrade mode. Once your iPhone is in DFU mode, redsn0w can take control of it (right).

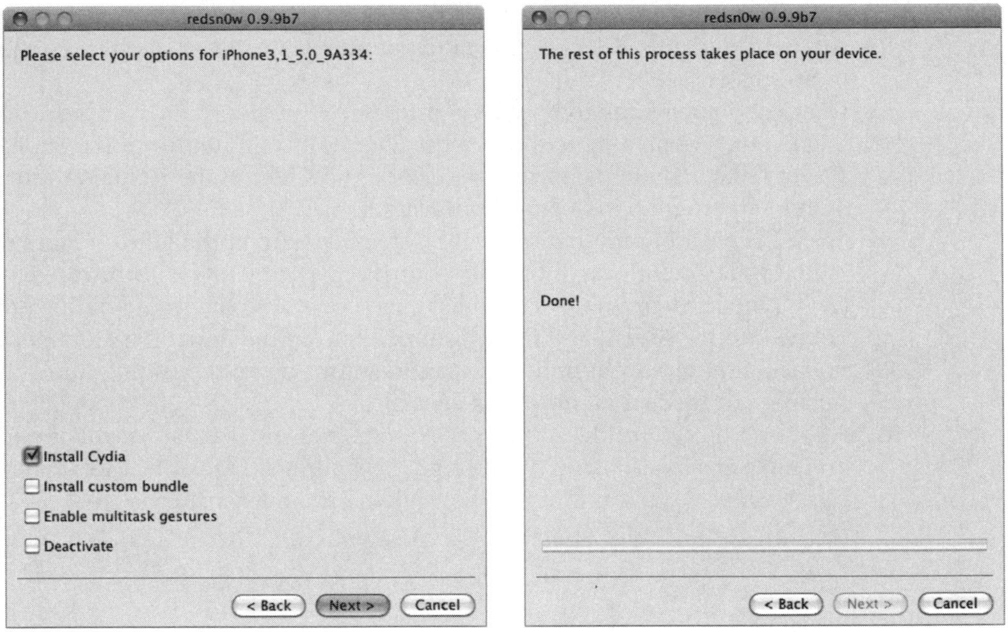

FIGURE 6-3   On the Please Select Your Options screen (left), select the Install Cydia check box, and then click the Next button. On the Done! screen (right), click the Cancel button to close redsn0w.

10. Select the Install Cydia check box.
11. Click the Next button. redsn0w installs Cydia and restarts your iPhone.
12. When you see the Done! screen (shown on the right in Figure 6-3), click the Cancel button to quit redsn0w.

Now that you've jailbroken your iPhone, you can unlock it from its carrier (as explained in Project 32 in Chapter 5) or install unapproved apps (as discussed in the next project.

Project 39: Find and Install Unapproved Apps

As you know, the official source for apps for your iPhone is Apple's App Store, which you can access either using iTunes on your computer or using the App Store app on your iPhone. The App Store has more than a half-million apps available at this writing, with more being added each day—so there's a wide variety you can choose from.

These are all apps that Apple has approved as being suitable for the iOS devices—the iPhone, the iPad, and the iPod touch.

To gain approval, an app not only must be programmed following Apple's guidelines but must also not violate any of its rules about content. For example, an app containing hardcore adult content won't get approved even if its coding is immaculate. Nor will an app that uses the underlying parts of iOS in ways that Apple doesn't permit, no matter how ingenious or useful the app is.

Because of this approval process, some developers choose not to submit their apps to the App Store. Instead, they make them available through other sources.

At this writing, Cydia is the main tool for installing unapproved apps on iOS devices. After you install it on a jailbroken iPhone, Cydia gives you access to a wide range of repositories for iOS software. This software includes both free apps and paid apps that you buy through the Cydia Store.

 This project assumes you've jailbroken your iPhone and installed Cydia as described in the previous project. If not, go back and do so. If you used a tethered jailbreak, use the jailbreaking software to boot into the jailbroken state.

Open Cydia

To open Cydia, tap the Cydia icon on one of your Home screens, just like any other app. In the left screen in Figure 6-4, you can see the Cydia icon in the lower-left corner, just above the Phone icon.

Tap the Cydia icon to open Cydia, just as you would launch any other app.

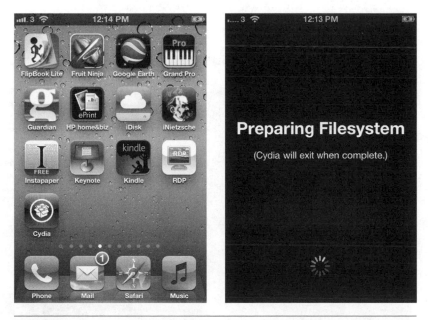

FIGURE 6-4 Tap the Cydia icon on one of your iPhone's Home screens (left) to launch Cydia. The first time you run Cydia, you'll see the Preparing Filesystem message (right).

DOUBLE GEEKERY

Understand Why Cydia Needs to "Prepare" the File System

Your iPhone obviously has a fully functional file system—if it didn't, it wouldn't be running. So you may well wonder why Cydia needs to "prepare" the file system.

What's happening here is that Cydia is moving apps and various other files from the OS partition to the Media partition and replacing them with symbolic links so that they'll continue to work. By moving the apps, Cydia frees up space on the OS partition, which enables you to put other apps on it.

We'll get into the details of the file system in Project 42.

The first time you run Cydia, you'll see the Preparing Filesystem message (shown on the right in Figure 6-4) for a few minutes while Cydia gets itself into shape. When Cydia finishes preparing the file system, it automatically quits.

Tap the Cydia icon to restart the app. Cydia displays the Who Are You? screen (shown on the left in Figure 6-5), which lets you choose your type of Cydia usage:

- **User** Tap this button to make apps, tweaks, and themes available. This is usually the best choice to start with.

FIGURE 6-5 On the Who Are You? screen (left), tap the User button, the Hacker button, or the Developer button, as appropriate, and then tap the Done button to display the Cydia screen (right).

- **Hacker**   Tap this button to make apps, tweaks, themes, and command-line tools available.
- **Developer**   Tap this button to make all the Cydia apps and utilities available.

After tapping the appropriate button, tap the Done button. You'll then see the Cydia app's interface, which consists of five screens, among which you switch by tapping the tabs at the bottom of the screen. The right screen in Figure 6-5 shows the Cydia screen, which you'll see at first.

Find Apps in Cydia

You can find apps in Cydia by using the Cydia screen, the Sections screen, the Changes screen, and the Search screen, which you access by tapping the tabs at the bottom of the screen.

- **Cydia**   On this screen, you can quickly access the Featured list, the Themes list, and the Cydia Store.
- **Sections**   Tap this tab to display a screen containing a list of different categories (sections) of apps and utilities, as shown on the left in Figure 6-6. Tap a category to display its contents, as shown on the right in Figure 6-6.

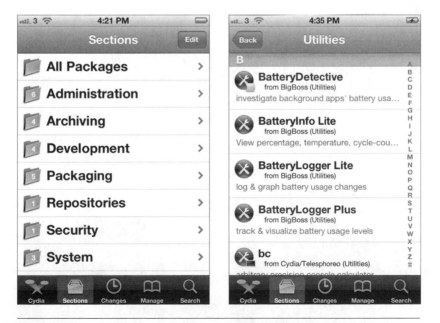

FIGURE 6-6   Use the Sections screen (left) to browse the available software by categories. Tap a category to display its contents (right).

In the Cydia listings, item names that appear in black are free. Item names that appear in blue are pay software. For pay software, you can pay using either Amazon Payments or PayPal.

- **Changes**   Tap this tab to display the Changes screen (shown on the left in Figure 6-7), which provides a list of the latest software.
- **Search**   Tap this tab to display the Search screen (shown on the right in Figure 6-7). You can then type a search term to find matches.

Install an App with Cydia

When you've found an app that interests you, tap its button to display the Details screen (shown on the left in Figure 6-8). You can then tap the Install button to install the app if it's free or the Purchase button to buy the app if it's not free.

On the Confirm screen that appears (shown on the right in Figure 6-8), tap the Confirm button to go ahead with the installation. You'll then see the installer run, as shown on the left in Figure 6-9.

When the installer displays the Complete screen, as shown on the right in Figure 6-9, tap the Return To Cydia button to close the installer and return to Cydia.

After installing some apps, you may need to restart Springboard, the iOS feature that runs the Home screen. If so, the installer displays a Restart Springboard button in place of the Return To Cydia button.

FIGURE 6-7   The Changes screen (left) lists the latest software. The Search screen (right) lets you search by keyword.

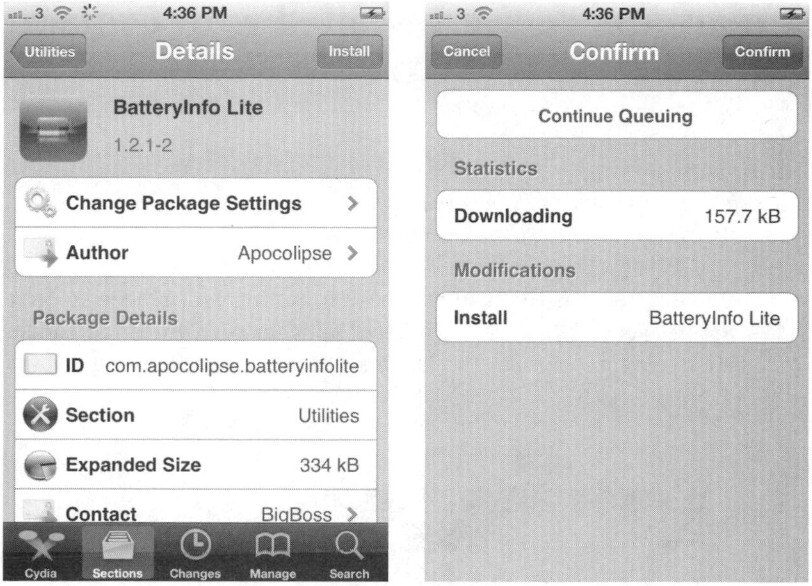

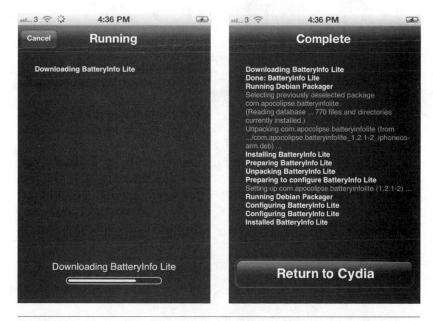

FIGURE 6-8   On the Details screen for an app (left), tap the Install button or the Purchase button. On the Confirm screen (right), tap the Confirm button.

FIGURE 6-9   The installer downloads the app's file and then installs it (left). When the installation process finishes, tap the Return To Cydia button (right).

Run an App You've Installed with Cydia

After installing an app with Cydia, the app appears on one of your iPhone's Home screens, just like when you install an app from the App Store. The left screen in Figure 6-10 shows the Home screen with several Cydia apps installed on the fourth line.

Tap the app's icon to open the app. The right screen in Figure 6-10 shows BatteryInfoLite, an app installed using Cydia.

 One difference is that you may need to restart your iPhone to get a freshly installed app to work. If you used a tethered jailbreak, you'll need to connect your iPhone to your computer and use the jailbreaking tool to perform the restart.

FIGURE 6-10 After you install an app using Cydia, its icon appears on the Home screen (left). Tap the icon to launch the app, and it runs as normal (right).

Uninstall an App You've Installed with Cydia

To uninstall an app you've installed with Cydia, follow these steps:

1. From the Home screen, tap the Cydia icon to launch Cydia.
2. Tap the Manage tab to display the Manage screen (shown on the left in Figure 6-11).
3. Tap the Packages button to display the Installed screen (shown on the right in Figure 6-11).
4. Tap the button for the app you want to remove. Cydia displays the Details screen for the app, as shown on the left in Figure 6-12.
5. Tap the Modify button. Cydia displays a dialog box, as shown on the right in Figure 6-12.
6. Tap the Remove button. Cydia displays the Confirm screen.
7. Tap the Confirm button. Cydia runs the installer, which uninstalls the app.
8. Tap the Return To Cydia button to return to Cydia.

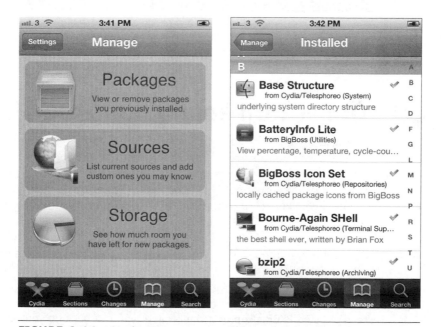

FIGURE 6-11 On the Manage screen (left), tap the Packages button to display the Installed screen (right), and then tap the button for the app you want to remove.

FIGURE 6-12   On the Details screen for the app (left), tap the Modify button, and then tap the Remove button in the dialog box that opens (right).

Project 40: Back Up Your Jailbroken iPhone

If you've followed through the previous two projects, you've now jailbroken your iPhone, installed some unapproved apps on it, and are enjoying using them.

Now for the bad news: If you update your iPhone's firmware to a new version, you may well lose the jailbroken apps. This is because iTunes doesn't include the folders that contain the jailbroken apps in the backup—so when it restores your iPhone after the firmware upgrade, those apps won't be there.

This doesn't mean you can't update your iPhone—it just means that you need to back up your jailbroken apps so that you can restore them after a firmware upgrade.

In this project, we'll use PKGBackup to back up your iPhone's jailbroken apps and to restore them. PKGBackup is a paid app that costs $7.99 at this writing.

 Instead of using PKGBackup or a similar app, you can back up your jailbroken files manually if you prefer. See Project 41 for instructions on connecting to your iPhone via Secure Shell from your computer and Project 42 for instructions on exploring your iPhone's file system to find the files you need to back up.

Buy and Install PKGBackup

To get PKGBackup, follow these general steps:

1. Run Cydia as explained in the previous project.
2. Search for PKGBackup, and then tap its button to display the Details screen.
3. Tap the Purchase button, and follow through the payment process. After your payment goes through, the Install button takes the place of the Purchase button.
4. Tap the Install button to install PKGBackup.
5. On the Confirm screen, tap the Confirm button.
6. When the Complete screen appears, tap the Return To Cydia button.
7. Press the Home button to return to the Home screen.

Run PKGBackup and Back Up Your Jailbroken Apps

After you install PKGBackup, run PKGBackup, choose settings, and back up your jailbroken apps. Follow these steps:

1. Tap the PKGBackup icon on the Home screen to launch the app.
2. If PKGBackup displays the Apps & Packages Scans Disabled dialog box, follow these steps to configure PKGBackup:
   a. Tap the Settings button in the Apps & Package Scans Disabled dialog box to display the PKGBackup screen in the Settings app. The left screen in Figure 6-13 shows the upper part of the PKGBackup screen in Settings, and the right screen in Figure 6-13 shows the lower part.

 If your iPhone displays the "PKGBackup" Would Like To Use Your Current Location dialog box, tap the Don't Allow button.

   b. In the At Startup box, set the Scan Applications switch, the Scan Packages switch, and the Automatic Backup switch to the On position.
   c. If you want to choose other settings, scroll down to the lower part of the PKGBackup screen (shown on the right in Figure 6-13), and choose them. For example, in the Dialogs box, you can choose whether to confirm backups, whether to confirm restores, and whether to enter a backup memo (a note about what a particular backup contains).
   d. When you finish choosing settings, tap the Settings button to display the Settings screen. Then press the Home button twice to display the app-switching bar, and tap PKGBackup to display the app again.
3. At this point, you should be seeing the PKGBackup screen (shown on the left in Figure 6-14). Tap the Settings button (the cog wheel icon) in the upper-left corner to display the Settings screen (shown on the right in Figure 6-14).

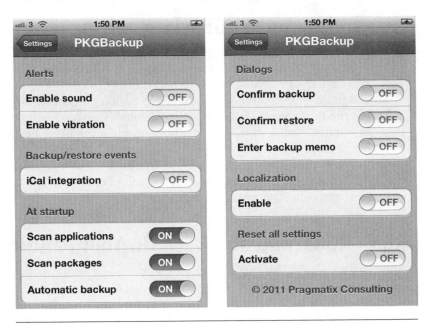

FIGURE 6-13 In the upper part of the PKGBackup screen in the Settings app (left), set the Scan Applications switch, the Scan Packages switch, and the Automatic Backup switch to the On position. You can also choose other settings on the lower part of the PKGBackup screen (right).

4. In the Select Where To Store Your Data box, tap the Connect To Dropbox button. PKGBackup displays the Link Account screen (shown on the left in Figure 6-15), which you use to link your Dropbox account so that PKGBackup can store data in it.

 If you don't have a Dropbox account yet, tap the Create An Account link at the bottom of the Link Account screen to start one.

5. Tap the Email box and type the e-mail address you use for your Dropbox account.
6. Tap the Password box and type the password for your Dropbox account.
7. Tap the Link button. PKGBackup establishes the link, and then displays the Settings screen again.
8. In the # Of Backups To Keep box, either enter a specific number of backups (for example, 5) or leave the default setting, 0, which allows an unlimited number of backups.

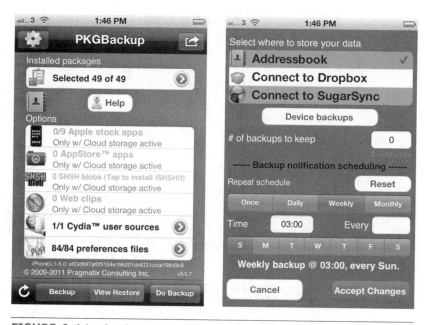

FIGURE 6-14   On the main PKGBackup screen (left), tap the Settings button (the cog wheel) to display the Settings screen (right).

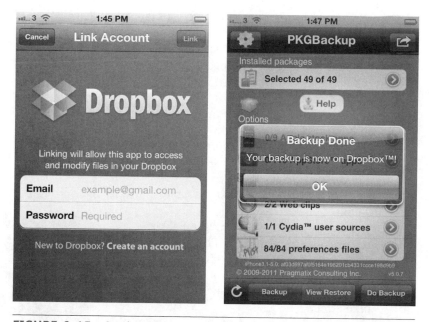

FIGURE 6-15   On the Link Account screen (left), enter the details of your Dropbox account or start creating a new account to link to PKGBackup. On the main PKGBackup screen, you can then tap the Do Backup button in the lower-right corner. When the Backup Done dialog box (right) appears, tap the OK button.

You may need to limit the number of backups to prevent PKGBackup loading your Dropbox account chock-full. But at first you may prefer to leave the 0 setting (for unlimited backups) until you see how much space each backup takes in Dropbox. You can then decide how many backups to keep, and enter that number in the # Of Backups To Keep box.

9. If you want to create a scheduled backup, use the controls in the Repeat Schedule section of the Settings screen to specify the details—for example, Daily at 03:00 or Weekly at 22:00 every Sunday.
10. Tap the Accept Changes button to save the changes you've made. PKGBackup returns you from the Settings screen to the main screen.
11. Tap the Do Backup button in the lower-right corner of the screen to run a backup now. You'll see a progress indicator as PKGBackup backs up your data. When PKGBackup displays the Backup Done dialog box, as shown on the right in Figure 6-15, tap the OK button.

Restore Your Jailbroken Apps with PKGBackup

When you need to restore your jailbroken apps, follow these steps:

1. Tap the PKGBackup icon on the Home screen to launch PKGBackup.

 If the reason you need to restore your jailbroken apps is that an iPhone firmware update has removed them, you will need to get PKGBackup up and running first. This means jailbreaking the iPhone, installing Cydia, using Cydia to install PKGBackup, and then connecting PKGBackup to your Dropbox account so that it can access your backups.

2. Tap the View Restore button at the bottom of the screen to display the Restore screen (shown on the left in Figure 6-16).
3. Tap the Select Backup button to display the list of available backups (shown on the right in Figure 6-16).
4. Tap the backup you want to use.
5. Tap the Select button. PKGBackup displays its main screen with the backup's details.
6. Tap the Do Restore button. PKGBackup restores the apps and then displays the Restore Done dialog box (shown here).
7. Tap the Reboot button if you're ready to restart your iPhone to make the changes take effect.

Select Backup
button

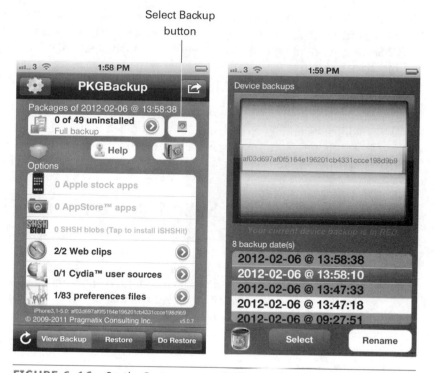

FIGURE 6-16   On the Restore screen (left), tap the Select Backup button to display the Device Backups screen (right). Tap the backup you want to use, and then tap the Select button.

Project 41: Connect to Your iPhone via SSH from Your Computer

In this project, we'll look at how to connect to your jailbroken iPhone using Secure Shell (SSH). Connecting via SSH enables you to access your iPhone's file system and transfer files back and forth.

 SSH is a networking protocol that you use to establish a secure connection between two computers. One computer is an SSH server, set up to accept connections from SSH clients. In this project, your iPhone is the SSH server and your computer is the SSH client.

Here's what we'll do in this project:

- Install the free SSH app called OpenSSH on your iPhone. This is the app that runs the SSH server on your iPhone.

- Install the free utility app called SBSettings on your iPhone. This app enables you to control Springboard settings and turn system services on and off. You need SBSettings to turn OpenSSH on and off, because OpenSSH doesn't have a user interface.
- Install the free SSH-capable application called FileZilla on your PC or Mac.
- Connect to your iPhone.

After you've established the connection, as discussed in this project, you can explore your iPhone's system partition and media partition (as discussed in Project 42) or recover space from your iPhone's system partition (as discussed in Project 43).

Install OpenSSH and SBSettings on Your iPhone

To install OpenSSH and SBSettings on your iPhone, follow these steps:

1. Run Cydia as explained in Project 39.
2. Search for **openssh**, and then tap its button to display the Details screen.
3. Tap the Install button. Your iPhone displays the Confirm screen.
4. Tap the Confirm button to confirm the installation. Cydia then downloads OpenSSH and runs the installer.
5. When the installer displays the Complete screen, tap the Return To Cydia button to return to Cydia.
6. Repeat steps 2–5 but this time search for sbsettings and install the SBSettings app.

Install FileZilla on Your Computer

Next, download FileZilla and install it on your PC or Mac. Follow these steps:

1. Open your web browser and go to the FileZilla website, http://filezilla-project.org.
2. Download and install the latest version of the FileZilla Client for Windows or for Mac, as appropriate.
   - **Windows** Run the file you download, and then follow through the setup routine. If you are an Administrator for your PC, you can choose whether to install FileZilla for all users or only for you. And on the Choose Components screen (shown next), choose which of the optional items to install. The Shell Extension component lets you drag files between Internet Explorer and FileZilla and is usually helpful; whether to install the Icon Sets, Language Files, and Desktop Icon components is up to you.

- **Mac OS X**   Open the compressed file if Safari doesn't open it for you, and then drag the FileZilla application to the Applications folder. Leave the Applications folder open for now so that you can open FileZilla in the next step.
3. Open FileZilla:
    - **Windows**   On the Completing The FileZilla Client Setup screen of the installer, select the Start FileZilla Now check box, and then click the Finish button. In the future, choose Start | All Programs | FileZilla FTP Client | FileZilla.
    - **Mac OS X**   In the Finder window showing the Applications folder, hold down OPTION and double-click the FileZilla icon. (Holding down OPTION as you double-click the icon makes the Finder window close as the application opens.)
4. If FileZilla displays the Welcome To FileZilla dialog box, click the OK button to close it. You'll then see the main FileZilla window. Figure 6-17 shows the Mac version.

Use SBSettings to Find Your iPhone's IP Address and Turn On SSH

Now run SBSettings and use it to find your iPhone's IP address and to turn on SSH. Follow these steps:

1. From the Home screen, tap the SBSettings icon to launch SBSettings.
2. Press the Home button to display the Home screen again.
3. Swipe your finger from left to right across the status bar at the top of the Home screen to display the SBSettings panel.

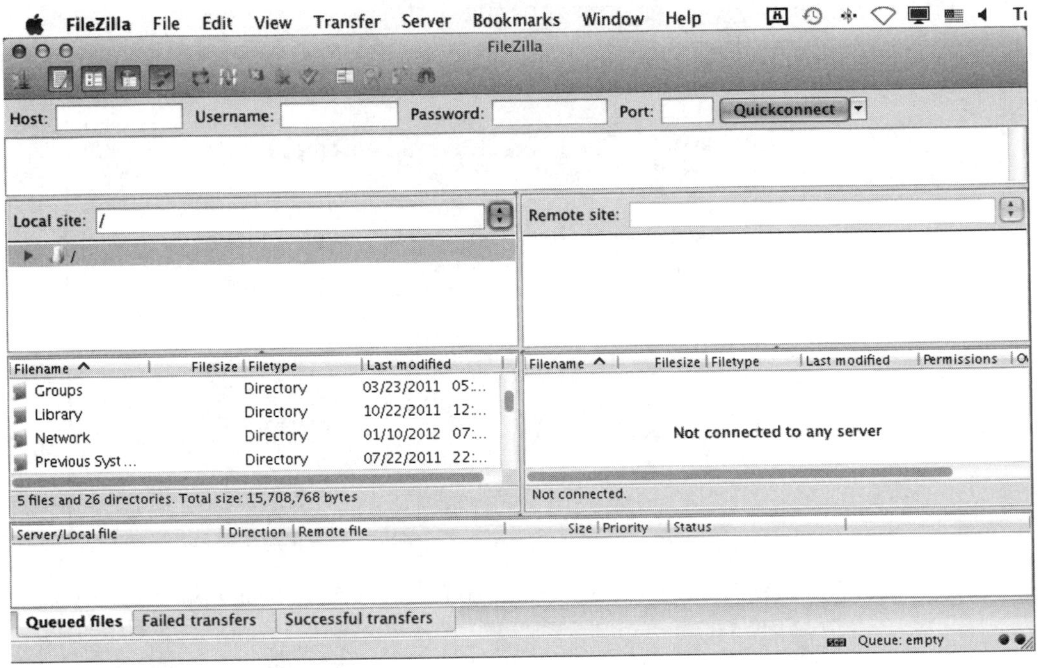

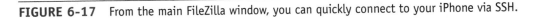

FIGURE 6-17 From the main FileZilla window, you can quickly connect to your iPhone via SSH.

4. Note the IP address shown in the Wi-Fi IP Address readout near the bottom—for example, 10.0.0.44 or 192.168.1.153.
5. If the SSH icon on the right side of the second line is red (indicating that SSH is off), tap the icon. When the icon turns green, SSH is on.
6. Tap the Close button (the × button) at the upper-left corner to close the SBSettings panel.

Create the Connection in FileZilla

Now create the connection to your iPhone in FileZilla. Follow these steps:

1. Click the Site Manager button at the left end of the toolbar, or choose File | Site Manager, to display the Site Manager dialog box (shown in Figure 6-18 with a site for the iPhone being created).
2. Click the New Site button. FileZilla creates a new entry in the My Sites list in the Select Entry pane and names it New Site.
3. Type the name for the site—for example, **My iPhone**—over the default name and press ENTER (on Windows) or RETURN (on the Mac) to apply the new name.
4. Click in the Host box and type the IP address you learned in the previous section.
5. Leave the Port box blank.
6. Open the Protocol drop-down list and choose SFTP – SSH File Transfer Protocol.
7. Open the Logon Type drop-down list and choose Normal.

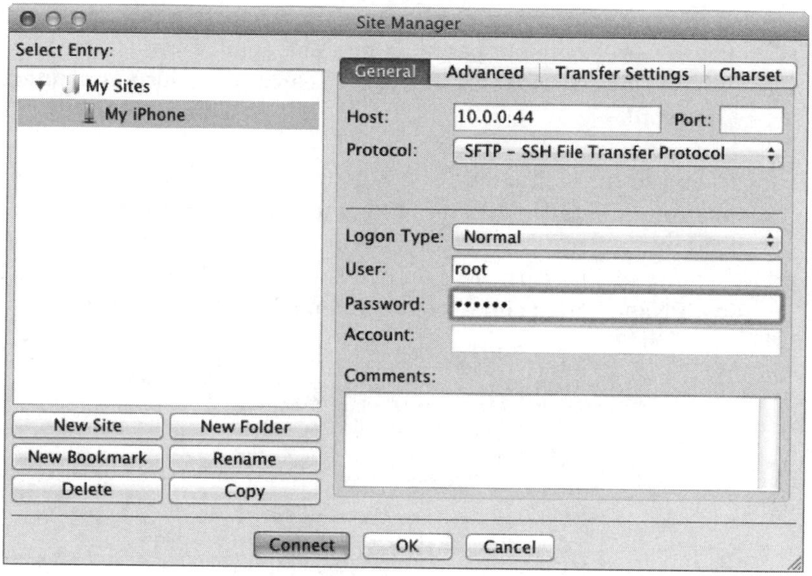

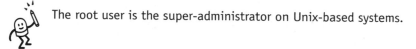

FIGURE 6-18   From FileZilla's Site Manager dialog box, you can create FTP sites, manage them, and connect to them.

8. Click in the User box and type **root**.

 The root user is the super-administrator on Unix-based systems.

9. Click in the Password box and type the standard password, **alpine**.

   Leave the Site Manager open so that you're ready to connect as described in the next section.

Connect to Your iPhone

Now that you've created a site for your iPhone in FileZilla, you can connect to it quickly. Follow these steps:

1. Make sure that your PC or Mac is connected to the same network as the iPhone.

 Your PC or Mac doesn't necessarily have to be connected to the same *wireless* network as the iPhone. If you have a network that combines wired and wireless portions, your computer can be connected to the wired portion and the iPhone can be connected to the wireless portion.

2. In the Site Manager window in FileZilla, click your iPhone's site, and then click the Connect button.

If the password **alpine** doesn't work for connecting to your iPhone, and you haven't set a different password by using a jailbroken utility, search online for other standard passwords to try. Use search terms such as **connect iphone ssh password.**

3. If you see the Unknown Host Key dialog box (shown here), which warns you that your computer doesn't know the SSH server's host key and so can't confirm its identify, verify the IP address on the Host line, and then click the OK button. If you're feeling trusting, you can select the Always Trust This Host, Add This Key To The Cache check box before clicking the OK button.

FileZilla then displays your iPhone's file system in the right pane, as shown in Figure 6-19. The left pane shows the current folder on your PC or Mac.

You're now ready to explore your iPhone's partitions. See the next project for details.

FIGURE 6-19 FileZilla displays your iPhone's file system in the right pane.

DOUBLE GEEKERY

Change Your iPhone's Root Password

When you read in the main text that most iPhones use the same root password, "alpine," did you think "Uh-oh..."?

As I'm sure you're all too well aware, a password known is a password blown. So if you want to be able to connect safely to your iPhone via SSH, you need to change your iPhone's root password.

Apple doesn't give you a way to do this, so you need to turn to jailbroken software. What you need is a terminal app called MobileTerminal—the iOS equivalent of Command Prompt (in Windows) or the Terminal utility (on the Mac), which you use to give the commands for changing the password.

Open Cydia, tap the Search tab at the bottom, and then search for **mobileterminal**. Tap the search result and read the description.

If the description says MobileTerminal is compatible with the version of iOS you're using (iOS 5 at this writing), tap the Install button to download and install it.

If the description says MobileTerminal isn't compatible with that version of iOS, you'll need to use the external repository at iJailbreak.com. Follow these steps to add this repository as a source for apps:

1. In Cydia, tap the Manage tab at the bottom to display the Manage screen.
2. Tap the Sources button to display the Sources screen (shown on the left in the next illustration).
3. Tap the Edit button in the upper-left corner to turn on Edit mode. The Add button replaces the Edit button.
4. Tap the Add button to display the Enter Cydia/APT URL dialog box (shown on the right in the illustration).

5. Type in the URL **http://www.ijailbreak.com/repository/**.
6. Tap the Add Source button. You'll see the Updating Sources screen as Cydia updates its list of sources.
7. When the Complete screen appears, tap the Return To Cydia button to return to Cydia.
8. Tap the Search tab and search for **mobileterminal** again. This time you'll see the version from the iJailbreak.com repository. This version is compatible with iOS 5.
9. Tap the Install button to display the Confirm screen.
10. Tap the Confirm button. Cydia downloads and installs MobileTerminal.
11. Tap the Return To Cydia button to return to Cydia.

You can now use MobileTerminal to change your password. Follow these steps:

1. On the Home screen, tap the Terminal icon to launch MobileTerminal (shown in the illustration with the sequence of commands being issued).
2. Type the following command:

   `su root`

3. Tap the Return button. MobileTerminal prompts you for the password.
4. Type the default password:

   `alpine`

5. Tap the Return button. You'll see another prompt, like this:

   `iPhone:/variable/mobile root#`

6. Type the command for changing the password:

   `passwd`

7. Tap the Return button. MobileTerminal prompts you to enter the new password.
8. Type the new password you want to use.
9. Tap the Return button. MobileTerminal prompts you to retype the new password.
10. Type the new password again, and tap the Return button again. You'll then see the prompt again:

    `iPhone:/variable/mobile root#`

11. Type the exit command:

    `exit`

12. Tap the Return button.
13. Press the Home button to return to the Home screen.

Your iPhone is now using the new root password you set. From now on, you will need to use this password to connect via SSH.

Project 42: Explore Your iPhone's OS Partition and Media Partition

Once you've connected to your iPhone via SSH, as described in the previous project, you're ready to explore its partitions. In this section, you learn about the two partitions, how to explore them, how to copy files to or from your iPhone, and how to disconnect from the iPhone when you finish.

 To follow this project, you must already have connected to your iPhone via SSH, as described in Project 41.

Understand the Two Partitions

Your iPhone uses two partitions, the OS partition and the Media partition.

OS Partition

The OS partition contains the files for iOS and other essential files.

This partition is relatively small—the size varies depending on the version of iOS, but for iOS 5 it is typically between 1GB and 2GB.

The OS partition is normally set to read only, and iOS is designed not to write to it. Under normal use, the only times the OS partition is written to is when you install firmware updates and when you restore the iPhone.

Cydia makes the OS partition readable so that it can make changes to it. Cydia makes space for itself, plus extra space for apps that need to be on the OS partition, by moving the Applications folder (which contains the built-in apps) and various other folders from the OS partition to the Media partition. Cydia creates a symbolic link to the Applications folder and the other folders so that the apps still run as usual and iOS works normally.

 A *symbolic link* or *symlink* is a file that refers to another file or folder, much like a shortcut in Windows or an alias on the Mac.

Media Partition

The Media partition contains your media files—songs, videos, podcasts, and so on. This partition takes up all the space left on your iPhone after the chunk taken by the OS partition.

For example, say you have a 64GB iPhone. Those 64 gigabytes are "marketing gigabytes" of a billion bytes each rather than true gigabytes of 1,073,741,824 bytes

---

DOUBLE GEEKERY

Understand Why Some Apps Must Run from the OS Partition

Most apps that are written using normal coding practices can run either from the OS partition (as Apple intends) or from the Media partition using symbolic links. After you jailbreak your iPhone and install Cydia, Cydia puts such apps on the Media partition, leaving space on the OS partition.

But some apps are *hard-coded*—they have the paths to the files they require written into the code, rather than using variables that point to where the files actually are. Hard-coded apps have to go on the OS partition, because they won't run correctly from the Media partition.

(1024 × 1024 × 1024 bytes), so the actual capacity is 59.6 true gigabytes. The OS partition takes up between 1GB and 2GB, leaving you with 57–58GB free on the Media partition.

To create space on the OS partition for itself and for any apps that can run only from the OS partition, Cydia moves various folders from the OS partition to the Media partition.

Meet the Partitions and Folders in Your iPhone's File System

After connecting to your iPhone's file system with FileZilla, you'll see the folders it contains. In this section, we'll take a quick tour through the key folders. This example uses Windows screens, but the moves are the same on the Mac.

To take the tour, follow these steps:

1. Set up the FileZilla window along the lines of Figure 6-20 so that you have a good view of the Remote Site pane:
   - Choose View | Message Log, removing the check mark from the menu item, to hide the message log. This is the pane that shows the commands—for example, Status: Directory Listing Successful.
   - Choose View | Transfer Queue, removing the check mark from the menu item, to hide the transfer queue. This is the pane at the bottom of the FileZilla window that shows the progress of file transfers.
   - If the Remote Directory Tree pane isn't displayed, choose View | Remote Directory Tree (placing a check mark next to the command) to display it.

- Drag the vertical bar between the Local Directory Tree pane and the Local Site pane (on the left) and the Remote Directory Tree pane and Remote Site pane so that the Remote Site pane is wide enough to show all its files. You may need to adjust the width of this pane while you're browsing. You may also need to change the width of the columns in the Remote Directory Tree pane by dragging the divisions between the column headings to the left or right.

2. At the top of the Remote Site pane, you'll see the root directory, represented by a forward slash (/) as is the custom on Unix-based file systems. Click the root directory to display a list of its contents in the Remote Directory Tree pane, as shown in Figure 6-20.

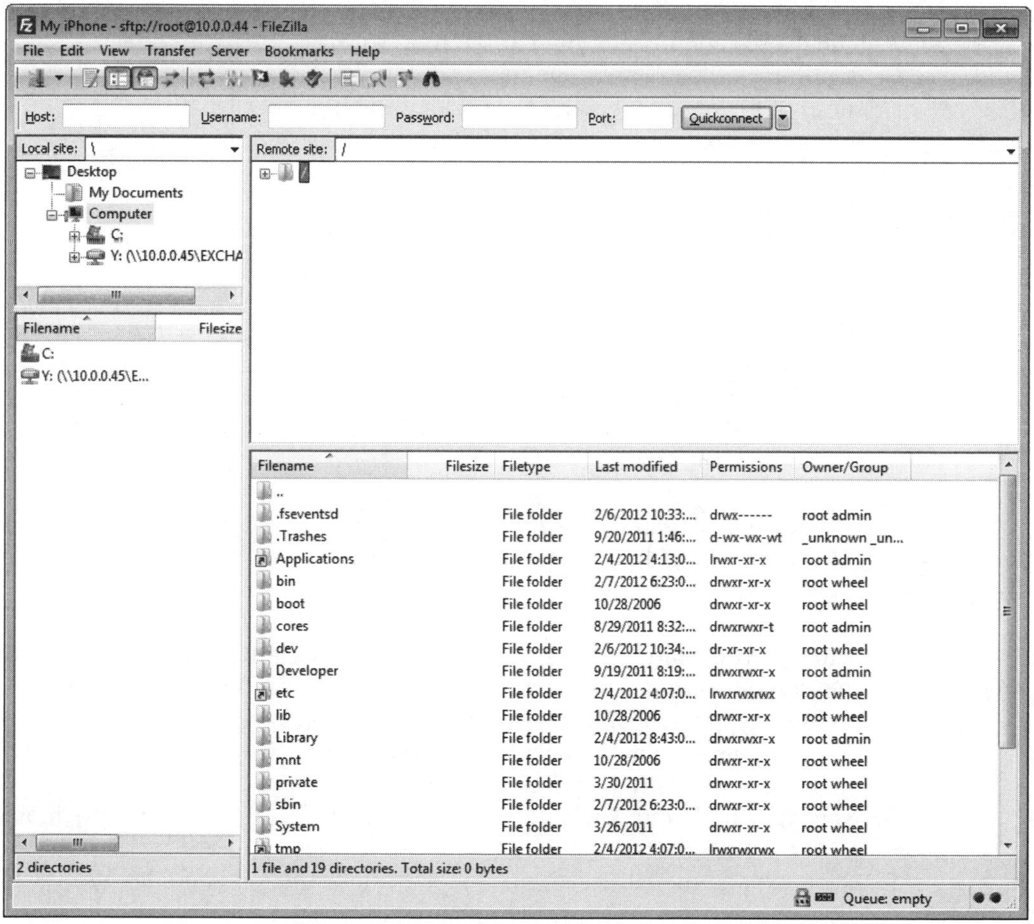

FIGURE 6-20   In the FileZilla window, hide the message log and the transfer queue, and then drag the main vertical divider bar to the left to give more space for the Remote Site pane (upper right) and the Remote Directory Tree pane (lower right).

3. If the root directory is collapsed, click the + sign or disclosure triangle to its left to expand it. You can also simply double-click the item. The illustration here shows the list of folders you'll see.

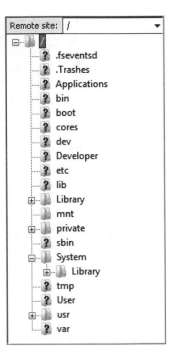

 As in Windows Explorer, a + sign in a box to the left of a folder in FileZilla indicates that you can expand it, and a – sign indicates that you can collapse it. Similarly, on the Mac, a gray downward-pointing disclosure triangle indicates that you can expand it, and a gray right-pointing disclosure triangle indicates that you can collapse it.

4. The OS partition is mounted at the root, so the contents of the OS partition appear directly inside the root folder—the Applications folder, the bin folder, the boot folder, and so on. The Media partition is mounted in the private folder, which we'll visit in a minute.

5. In the iPhone's normal, non-jailbroken state, the Applications folder contains the apps—Safari, Mail, Phone, and all the others. But as you read earlier, Cydia moves the contents of the Applications folder to give itself space on the OS partition. Try double-clicking the Applications folder. Instead of displaying the folder's contents, FileZilla follows the symlink and displays the contents of the /private/var/ stash/Applications.goGsbl folder (see Figure 6-21), which is where Cydia has moved the Applications folder.

 In Figure 6-21, you can also see the other items that Cydia has moved, including the Ringtones folder and the Wallpaper folder.

6. With the Applications.goGsbl folder selected in the Remote Site pane, look at the Remote Directory Tree pane. Here, you can see the list of apps in the folder, including AppStore.app, Camera.app, and Cydia.app.

7. Scroll up the Remote Site pane until you can see the mobile folder (still under /private/var/), and then double-click it to expand it.

8. Now, let's find your songs. First, expand the Media folder under the mobile folder.

9. Next, expand the iTunes_Control folder under the Media folder.

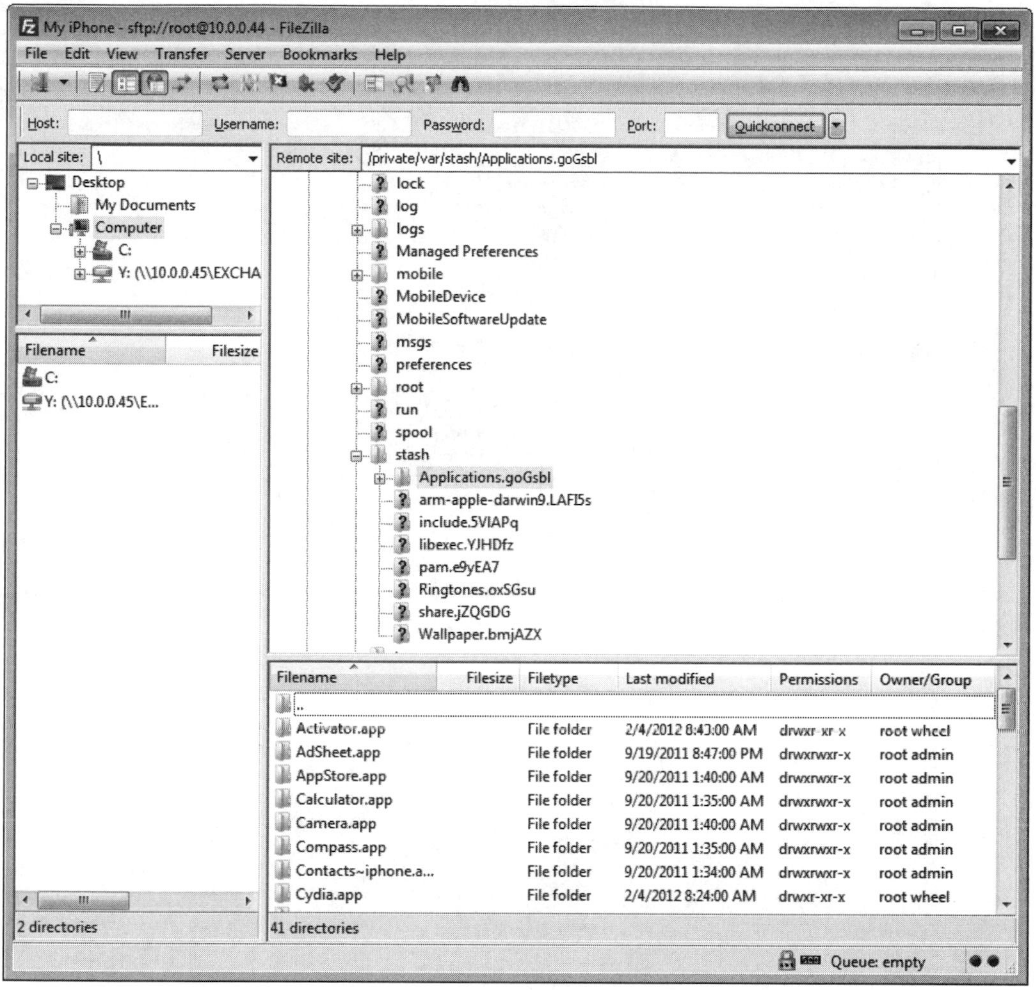

FIGURE 6-21   Double-clicking the Applications folder in a jailbroken iPhone takes you to the /private/var/stash/Applications.goGsbl folder, where Cydia has stored the apps.

10. Then expand the Music folder under the iTunes_Control folder.

11. Last, click one of the folders whose names begin with F—for example, the F00 folder. The list of song files it contains appears in the Remote Directory Tree pane (see Figure 6-22).

Looking at the songs listed in the Remote Directory Tree pane, you'll notice that they have cryptic, four-character names—for example, AATV.mp3 and AIKK.m4a. iTunes and your iPhone's Music app use these filenames, instead of the songs' titles (or mutations of them), to identify songs on the iPhone uniquely.

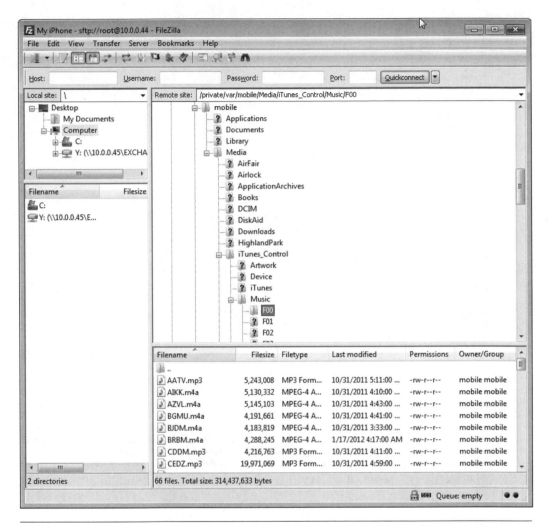

FIGURE 6-22 Open a subfolder of the /private/var/mobile/Media/iTunes_Control/Music/ folder to see the songs you've loaded on your iPhone.

As you can see, your iPhone has many other folders, but we'll stop the tour there for now. Leave the FileZilla window open if you want to copy or move files to or from your iPhone, as described next.

Copy Files to and from Your iPhone

After connecting to your iPhone with FileZilla, you can easily copy files to it or from it by dragging them between the Local Site pane and the Remote Site pane.

For storing your files on your iPhone, you'll probably want to create one or more of your own folders rather than using the iPhone's existing folders. To create a folder, follow these steps:

1. In the Remote Site pane, right-click (or CTRL-click on the Mac) the folder in which you want to create the new folder, and then click Create Directory on the context menu. FileZilla displays the Create Directory dialog box (shown here).

2. In the Please Enter The Name Of The Directory Which Should Be Created box, type the folder name over the New Directory placeholder.

Create folders only on the Media partition, not on the OS partition.

3. Click the OK button.

Disconnect FileZilla from Your iPhone

If you want to recover extra space from your iPhone's system partition, leave FileZilla open and move right on to the next project. If you're done with FTP-ing to your iPhone for now, click the Disconnect button (the button with the red ×) on the toolbar to disconnect from your iPhone.

Project 43: Recover Extra Space from Your iPhone's OS Partition

As you saw in the previous project, your iPhone uses two partitions: an OS partition that's mostly filled with essential files for iOS and the built-in apps, and a Media partition intended for your music, video, and other files.

If you're running out of space on your iPhone, you may want to recover free space from the OS partition so that you can make the Media partition that much larger.

In this project, I'll show you how to find out how much free space the OS partition has so that you can decide whether you want to recover it, as there may not be enough for this move to be worthwhile. I'll then show you how to recover the space by performing an untethered jailbreak using PwnageTool on the Mac.

 At this writing, Sn0wbreeze, the Windows equivalent of PwnageTool, can perform only a tethered jailbreak, not an untethered one. When you read this, check to see if the current version of Sn0wbreeze can perform an untethered jailbreak. If so, download Sn0wbreeze, extract the application file from the zip file, and double-click the application file to run Sn0wbreeze. You'll then need to go through User Account Control on Windows 7 or Windows Vista to allow Sn0wbreeze to run. After that, jailbreaking on Sn0wbreeze works in the same way as jailbreaking on PwnageTool, so you can follow the instructions in this section.

See How Much Space Is Free on the OS Partition

To see how much space is free on your iPhone's OS partition, follow these steps:

1. From the Home screen, launch Cydia.
2. Tap the Manage tab to display the Manage screen (shown on the left in Figure 6-23).

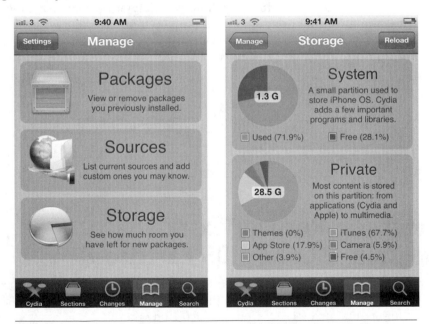

FIGURE 6-23 Tap the Storage button on the Manage screen (left) to display the Storage screen (right), which shows how much space is free on your iPhone's System partition (the OS partition) and on the Private partition (the Media partition).

3. Tap the Storage button to display the Storage screen (shown on the right in Figure 6-23).
4. In the System box at the top, look at the Used readout and the Free readout. (The System partition is the OS partition.)

The Private box at the bottom of the Storage screen shows a breakdown of what's taking up space on your iPhone—Themes, App Store, Other, iTunes, and Camera—and how much space is free.

Free Up Space on the OS Partition

If you decide there's enough space on the OS partition to be worth freeing up, and you use a Mac, you can perform an untethered jailbreak using the jailbreaking tool named PwnageTool. PwnageTool enables you to create a custom iOS firmware file that you then install by restoring it to your iPhone. While creating the custom firmware file, you can set the OS partition size.

To get PwnageTool, go to a site such as iJailbreak.com (www.ijailbreak.com) or Redmond Pie (www.redmondpie.com), follow the link to the Downloads section, and then download the latest version. If these sites no longer exist, search for PwnageTool using your favorite search engine or P2P client.

After downloading PwnageTool, open the disk image file if Mac OS X doesn't automatically open it for you. You can then either drag the PwnageTool icon to the Applications folder in the sidebar to install PwnageTool in the Applications folder, and then double-click its icon to run it, or simply double-click the icon to run PwnageTool from the disk image.

If PwnageTool displays the Check For Updates On Startup dialog box (shown here), click the Yes button to check for a newer version. Because Apple frequently changes iOS and the jailbreak developers must update their software to handle changes, it's a good idea to have the latest version.

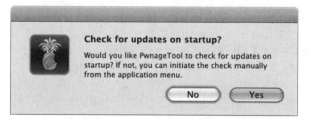

You then see the PwnageTool window (see Figure 6-24), and you can start using PwnageTool. Follow these steps:

1. Click the Expert Mode button on the toolbar to switch from Simple mode to Expert mode.
2. In the main part of the window, click the iPhone entry.
3. In the lower-right corner of the window, click the arrow button. PwnageTool displays the Browse For IPSW screen (see Figure 6-25).

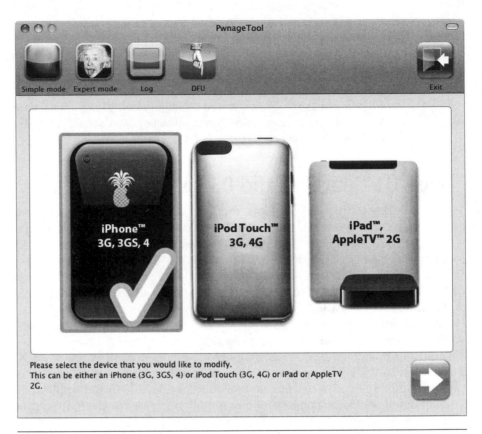

FIGURE 6-24 In the PwnageTool window, click the Expert Mode button on the toolbar, click the iPhone entry in the main pane, and then click the arrow button in the lower-right corner.

 IPSW is the abbreviation and file extension for iPod Software files. (The name is "iPod Software" rather than "iPhone Software" largely because the iPod preceded the iPhone.) An IPSW file is a Zip archive containing the files needed to restore or update an iPhone's, iPad's, or iPod's firmware.

4. Double-click the Browse For IPSW button to display the Open dialog box.

 If you've downloaded the IPSW file, it'll be in your Downloads folder unless you put it somewhere else—in which case, you'll know where to look. If you want to use the latest IPSW file that iTunes has downloaded, you'll find it in the ~/Library/iTunes/iPhone Software Updates/ folder (where the tilde, ~, represents your home folder). Each time you download a new IPSW file, iTunes deletes the last one, so as not to eat up your hard drive—so if you want to keep it, make a copy of it before downloading an update.

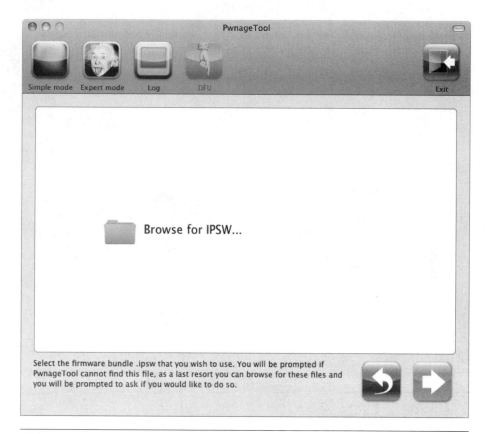

FIGURE 6-25   On the Browse For IPSW screen, double-click the Browse For IPSW button and then locate the iPhone software restore package you want to use.

5. Navigate to the folder that contains the IPSW file.
6. Click the IPSW file, and then click the Open button. PwnageTool displays the screen shown in Figure 6-26.
7. Click the General button to select it.
8. Click the arrow button in the lower-right corner to display the General Settings screen (see Figure 6-27).
9. In the Root Partition Size box, enter the size you want for the OS partition. You can either type in the number or drag the slider; usually, typing is easier.

 To work out how much space the OS partition needs, calculate your current usage from the Storage screen in Cydia, shown in Figure 6-23, earlier in this chapter.

10. Click the Back button—the button with the arrow curling counterclockwise—to return to the screen shown in Figure 6-26.

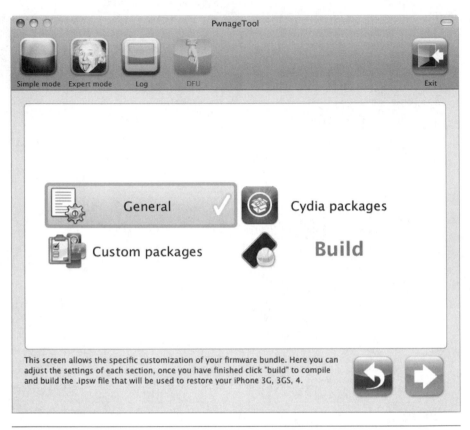

FIGURE 6-26 From this screen in PwnageTool, you can start assembling a custom IPSW file for your iPhone. You can also choose the size for the OS partition.

11. If you want to control which Cydia packages the IPSW file contains, click the Cydia Packages button to select it, click the arrow button, and then make your choices on the Cydia Settings screen. Click the Back button when you're ready to move on.

12. If you want to add custom packages to the IPSW file, click the Custom Packages button to select it, click the arrow button, and then make your choices on the Custom Packages Settings screen. Click the Back button when you're ready to return to the configuration screen.

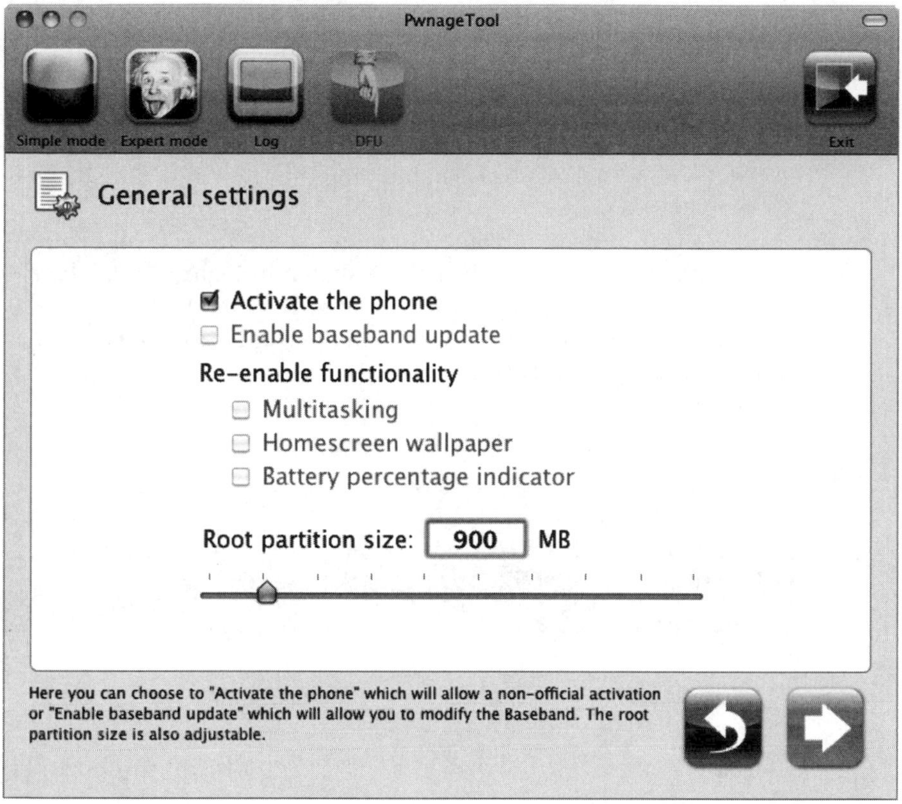

FIGURE 6-27 On the General Settings screen in PwnageTool, you can set the size of the root partition (the OS partition).

13. Click the Build button to select it, and then click the arrow button. PwnageTool displays the Save dialog box (shown here).

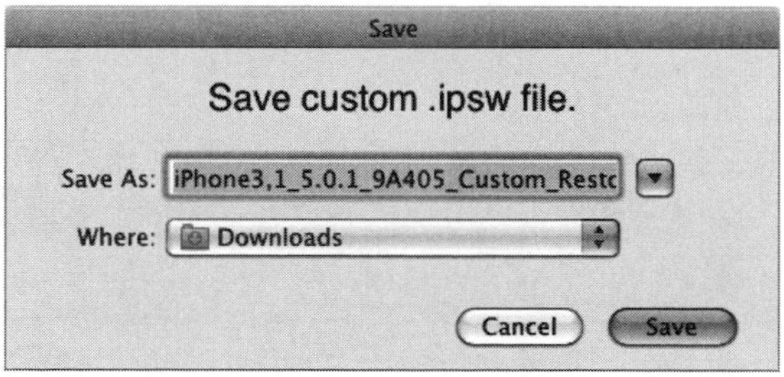

14. In the Save As box, edit the suggested name or type a new name as needed.

15. Choose where to save the file. You can choose a different folder in the Where pop-up menu, or click the down-arrow button to the right of the Save As box to expand the Save dialog box, and then navigate to the folder you want.
16. Click the Save button to save the file. PwnageTool builds the IPSW file and saves it using the folder and filename you chose.
17. When PwnageTool has finished building the IPSW file, it displays the Connect Device To USB screen (see Figure 6-28).
18. Connect your iPhone to your computer via the USB cable. PwnageTool detects your iPhone and displays the screen shown in Figure 6-29, which walks you through the steps to put your iPhone into DFU mode.
19. Follow the instructions for pressing the power button and Home button to put your iPhone into DFU mode.

FIGURE 6-28 When PwnageTool displays the Connect Device To USB screen, connect your iPhone to your computer via USB.

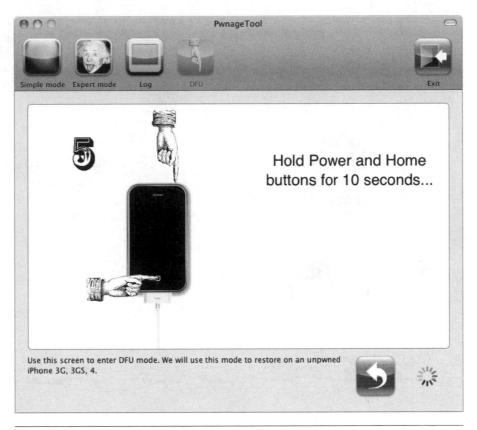

FIGURE 6-29 Follow the instructions on this PwnageTool screen to put your iPhone into DFU mode.

20. When PwnageTool displays the Successfully Entered DFU Mode! dialog box (see Figure 6-30), click the OK button. Then click the Exit button at the right end of the toolbar to exit PwnageTool.
21. iTunes then detects the iPhone as being in recovery mode, as shown in the illustration.

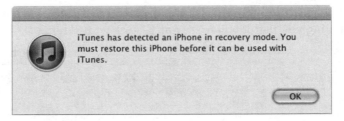

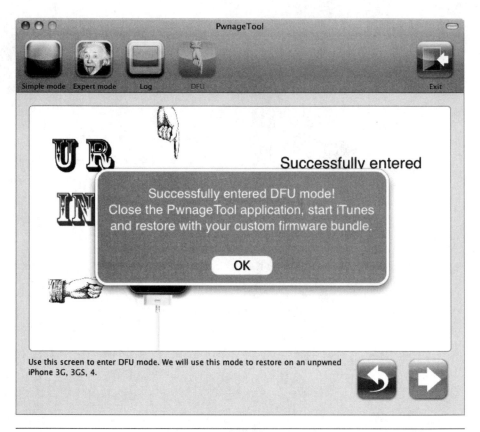

FIGURE 6-30 Click the OK button in the Successfully Entered DFU Mode! dialog box, and then click the Exit button at the right end of the toolbar to exit PwnageTool.

22. Click the OK button to close the dialog box. iTunes then adds the iPhone to the Devices category in the Source list and displays the Summary screen for the iPhone (see Figure 6-31).
23. OPTION-click the Restore button to display the Open dialog box.
24. Navigate to the folder in which you stored your custom IPSW file, and then click the file.
25. Click the Open button. iTunes displays a dialog box such as the one shown here, telling you that it will erase and restore your iPhone.

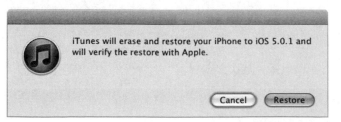

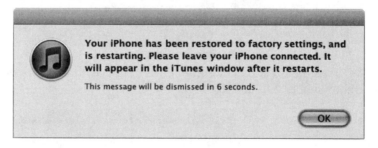

FIGURE 6-31   From the Summary screen of the iPhone's control screens, you can choose to restore your custom IPSW file to your iPhone.

26. Click the Restore button. iTunes proceeds with the restore operation, extracting the software from the IPSW file and then installing it on the iPhone. When iTunes finishes, you'll see the dialog box shown next, telling you that the iPhone has been restored to factory settings and is restarting.

> **Your iPhone has been restored to factory settings, and is restarting. Please leave your iPhone connected. It will appear in the iTunes window after it restarts.**
>
> This message will be dismissed in 6 seconds.
>
> OK

27. Either click the OK button to close the dialog box or allow iTunes to complete the countdown and close the dialog box itself.
28. Your iPhone then restarts and appears in the Devices list in iTunes. iTunes displays the Set Up Your iPhone screen (see Figure 6-32).
29. Select the Restore From The Backup Of option button.
30. In the pop-up menu, choose the correct iPhone and backup. The latest backup has just the iPhone's name; earlier backups show their dates and times as well.

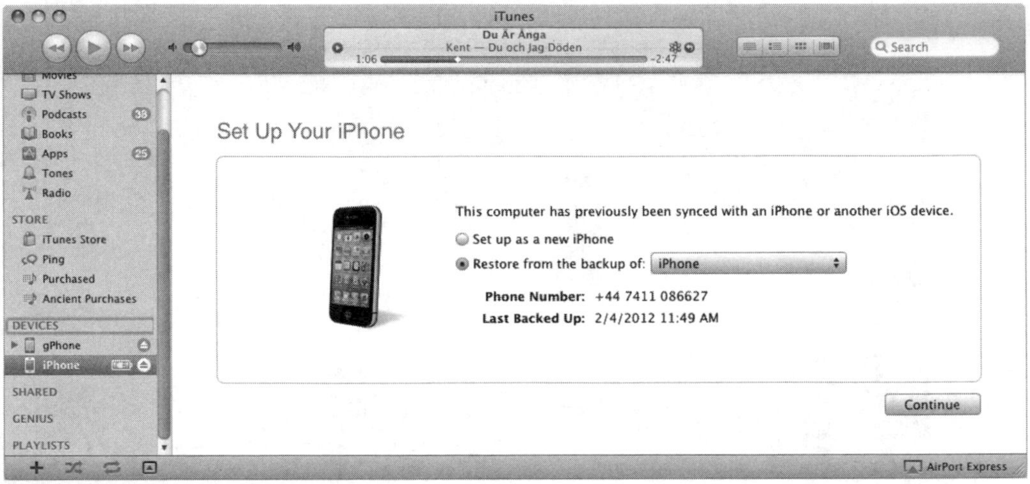

FIGURE 6-32 On the Set Up Your iPhone screen, select the Restore From The Backup Of option button, choose the correct iPhone and backup in the pop-up menu, and then click the Continue button.

31. Click the Continue button. If you encrypted the backup with a password, iTunes displays the Enter Password dialog box (shown here); type the password and press RETURN or click the OK button.

32. iTunes then restores the iPhone from the backup, displaying its progress (as shown here) while it does so.

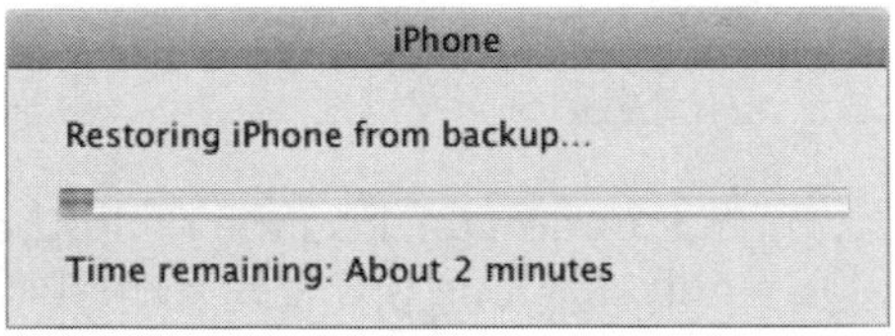

33. iTunes restarts your iPhone, displays all the tabs of its control screens, and starts syncing the apps and other content to it.

When the sync finishes, open Cydia, tap the Manage tab, and then tap the Storage button to see the new sizes of your iPhone's partitions.

Project 44: Apply a Theme to Your iPhone

If you want to make your iPhone's user interface look different, you can apply a theme to it. A *theme* is a different look—wallpaper, icons, and so on.

You can either download a theme using Cydia or find themes on the Web and then apply them yourself.

Install a Theme Using Cydia

To install a theme using Cydia, follow these steps:

1. If Cydia isn't running, tap the Cydia button on the Home screen to launch it.
2. Tap the Sections tab to display the Sections screen.

 You can also search for themes. For example, tap the Search tab, and then type **theme** on the Search screen—or, if you're looking for a theme whose name you know, type a distinctive word in the name.

3. Scroll down to the Themes part of the list. You'll find a large number of different items here—Themes, Themes (Carrier), Themes (Complete), Themes (System), and so on.
4. Tap the category of themes you want to browse.
5. Tap the theme you want to view. The Details screen appears.
6. If you want to install the theme, tap the Install button. The Confirm screen appears.
7. Tap the Confirm button. Cydia launches the installer, which downloads the theme and installs it.
8. Tap the Return To Cydia button to return to Cydia.

 Most themes you install by using Cydia include the WinterBoard app for choosing the theme. If the app you choose doesn't include WinterBoard, or if you download a theme and install it manually, search for **winterboard** in Cydia and install it yourself.

Install a Theme Manually

Installing a theme using Cydia is handy, but you'll find other themes on the Web that aren't available in Cydia packages. You need to install such themes manually. Follow these steps:

1. Download the theme to your computer.
2. Unzip the Zip file that contains the theme. You'll get a folder containing the files for the theme.

3. Connect to your iPhone using FileZilla, as explained in Project 41.
4. Copy the folder containing the theme's files to the /var/stash/Themes/ folder.

You can now apply the theme using WinterBoard, as described in the next section.

Apply a Theme Using WinterBoard

After installing a theme using Cydia (or installing it manually), use WinterBoard to apply the theme. Follow these steps:

1. Press the Home button to display the Home screen.
2. Tap the WinterBoard button to launch the WinterBoard app (shown on the left in Figure 6-33).
3. Tap the Select Themes button to display the screen shown on the right in Figure 6-33.
4. Tap to place a check mark on each item you want to use.
5. Tap the WinterBoard button in the upper-left corner to return to the WinterBoard screen.
6. Tap the Respring button in the upper-left corner to relaunch Springboard. You'll then see the theme, as shown in Figure 6-34.

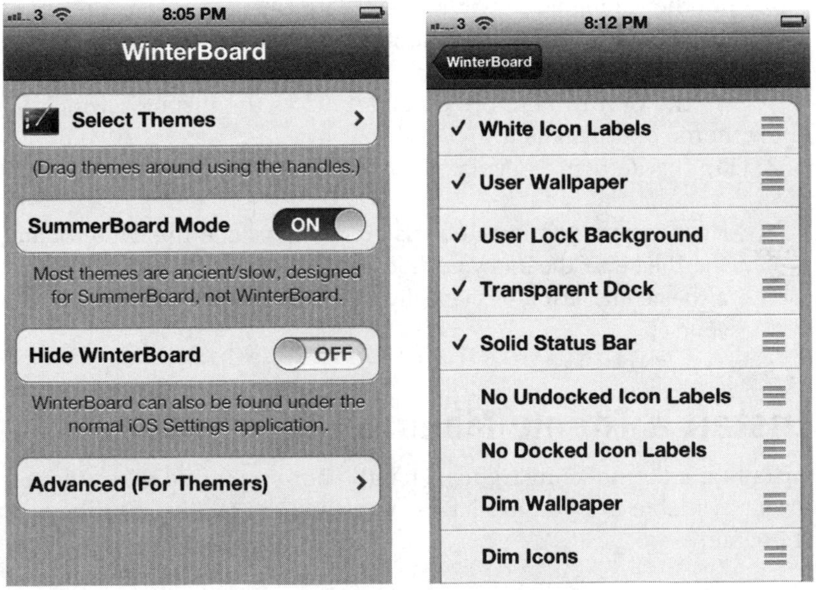

FIGURE 6-33 On the WinterBoard screen (left), tap the Select Themes button to display the screen for choosing the theme elements (right). Then tap to place a check mark on each item you want to use.

FIGURE 6-34   Your chosen theme appears after you relaunch Springboard.

Project 45: Make Wi-Fi–Only Apps Run over a 3G Connection

Some apps are designed to run only over Wi-Fi connections rather than over both Wi-Fi connections and 3G connections. The usual reason for this is that the app typically transfers large enough amounts of data to run through a normal data allowance uncomfortably fast.

But if you have a generous data plan, or if the app is so vital that you're prepared to pay any extra costs you run up, you may want to run an app over 3G.

To do this, you need an app called My3G. My3G costs $3.99 from the Cydia Store, but there's a three-day free trial that you'll probably want to test first to see if the app suits you.

Get and Install My3G

Run Cydia, tap the Search tab to display the Search screen, and then search for **my3g**. Tap the search result to display the Details screen, tap the Install button, and then tap the Confirm button. When the Complete screen appears, tap the Restart Springboard button to restart Springboard.

Tap the My3G icon on the Home screen to launch My3G. If you're using the trial version, tap the Start Trial button on the Welcome To My3G screen. My3G then downloads a trial license, after which you have to restart Springboard.

The easiest way to restart Springboard is to display SBSettings by dragging your finger across the status bar on the Home screen, and then tap the Respring button.

Now tap the My3G icon on the Home screen to launch My3G again, and you'll be in business.

Specify Which Apps to Run over 3G

What you need to do now is specify which apps you want to run over 3G. Normally, you'll pick only certain apps rather than letting the whole herd of apps run hog-wild through your data allowance.

On the My3G screen (shown on the left in Figure 6-35), tap the button for each app you want to use, putting a check mark next to it.

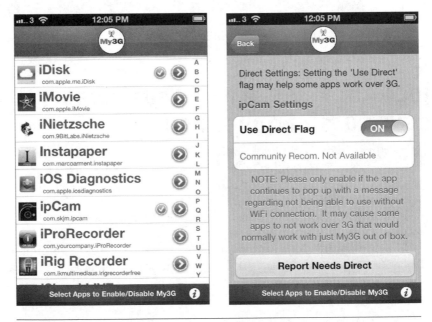

FIGURE 6-35   On the My3G screen (left), tap to place a check mark on each app you want to use over 3G. If you find an app displays messages warning that it requires Wi-Fi, tap the > button on the app's button on the My3G screen to display the Settings screen (right), and then move the Use Direct Flag switch to the On position.

Run Your Apps over 3G

You can now run those apps you chose over the 3G network instead of a wireless network. If you try to run an app, and it gives a message saying it requires Wi-Fi, turn on the direct flag for it like this:

1. Go back to the My3G app. For example, press the Home button twice in quick succession, then tap the My3G icon on the app-switching bar.
2. Tap the > button on the app's button to display the Settings screen (shown on the right in Figure 6-35).
3. Tap the Use Direct Flag switch and move it to the On position.

Now try the app again, and it should work over 3G.

Project 46: Play Games Under Emulation

You can find a vast number of games at the App Store—but there are many older games that people want to play: Sega Genesis, Nintendo and Super Nintendo, Game Boy Advance, PlayStation...and even old arcade games.

To play games that aren't designed to run on the iPhone, you need to install and use an emulator.

This project shows you how to get an emulator up and running, how to install games on the emulator, and how to run the games.

Install Your Emulator and Get It Running

Your first step is to get a suitable emulator and get it running. Here's the list of the emulators you'll need:

Game Console	Emulator	Cost
Sega Genesis	genesis4iphone	Free
Game Boy Advance	gpSPhone	$4.99
Arcade games	mame4iphone	Free
Nintendo	NES	$5.99
Super Nintendo	snes4iphone	Free
PlayStation	psx4iphone	$2.99

To get one of these emulators, follow these general steps:

1. Open Cydia by tapping its icon on the Home screen.
2. Tap the Search tab to display the Search screen.

3. Search for the emulator by name.
4. Tap the search result to display the Details screen.
5. Tap the Install button. Cydia displays the Confirm screen.
6. Tap the Confirm button. Cydia launches the installer, which downloads and installs the app.
7. Tap the Return To Cydia button or the Relaunch Springboard button.

Install Games on the Emulator

To install a game, connect to your iPhone using FileZilla, as described in Project 41, earlier in this chapter. Then use FileZilla to copy the game's ROM to the appropriate folder on your iPhone:

- **Sega Genesis**   /var/mobile/Media/ROMs/GENESIS/

 A ROM is a read-only memory file containing the game. If you don't have the ROMs for the game consoles you want to install, you can almost certainly find them on the Internet. It's a good idea to check that whoever is distributing them is doing so legally.

- **Game Boy Advance**   /var/mobile/Media/ROMs/GBA/

 For Game Boy Advance, you must also install a file named gba_bios.bin in the /var/mobile/Media/ROMs/GBA/ folder. You can find this file by searching online.

- **Multiple Arcade Machine Emulator**   /var/mobile/Media/ROMs/MAME/roms/
- **Nintendo**   /var/mobile/Media/ROMs/NES/
- **Super Nintendo**   /var/mobile/Media/ROMs/SNES/
- **Sony PlayStation**   /var/mobile/Media/ROMs/PSX/

 For PlayStation, you must also install a file named scph1001.bin in the /var/mobile/Media/ROMs/PSX/ folder. You can find this file by searching online. If you can't find the file by searching with Google, try Yahoo!.

Run the Games on the Emulator

After installing the games, you're ready to run them. Launch the emulator from the Home screen, pick the game from the list of those you've installed, and then start playing. The left screen in Figure 6-36 shows the list of games in the genesis4iphone emulator. The right screen in Figure 6-36 shows Mortal Kombat 3 ready for action.

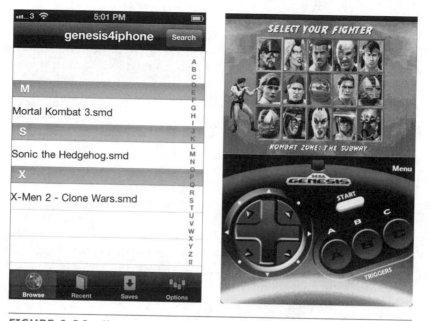

FIGURE 6-36 Choose the game in the emulator (left) and then start playing (right).

Project 47: Open Your iPhone and Eye Its Insides

Unlike some early iPhones, which were the devil's own job to open (and not much easier to close again), the iPhone 4S and iPhone 4 are a breeze to open. Once you've got the right kind of screwdriver, you'll have the back off your iPhone in two shakes of a lamb's tail.

Check Which Kind of Screws Your iPhone Uses

First, check which kind of screws your iPhone uses. Turn it upside down, as shown here, and look at the screws on either side of the Dock Connector port at the bottom:

- **Phillips screws** If the iPhone's screws have a cross-shaped slot, all you need is a regular Phillips screwdriver, size #00. Many iPhone 4 models use these screws.

- **Pentalobe screws**   If the iPhone's screws have a five-pointed star at the top, you need a pentalobe screwdriver. You can find these anywhere online for a couple of dollars. For example, go to eBay and search for **pentalobe screwdriver** or **iphone opening toolkit**. Later iPhone 4 models and most iPhone 4S models use pentalobe screws.

Take Off the Back of the iPhone

Armed with your Phillips screwdriver or pentalobe screwdriver, you can take off the back of your iPhone like this:

1. Turn the iPhone off to reduce the chance of electrical mishaps. Hold down the Sleep/Wake button until the Slide To Power Off slider appears, and then slide the slider.
2. Place the iPhone face down on a soft cloth on a flat surface.
3. Discharge any static electricity you're carrying by touching a convenient metal object (not your iPhone).
4. Unscrew the two screws at the bottom of the iPhone, as shown in the illustration. Put the screws somewhere safe—they're easy to lose.

5. Slide the back panel up toward the top of the iPhone, as shown in the illustration. It'll move only about 1/16th of an inch (2mm), but this is all that's needed to disengage the catches.

6. Lift up the back panel. Usually your fingernails will do the trick, but you may prefer to use a plastic spudger instead, as shown in the illustration.

7. Take the back panel off.

Identify the iPhone's Components

After taking the back panel off, you can see many of the iPhone's key components, as shown in Figure 6-37. To get at the other components, you need to dig the battery out and then remove the EMI shield and the antenna/speaker enclosure.

Put the Back Plate on Again

Unless you're planning surgery on your iPhone, you should probably put the back plate on at this stage—or replace it with another back plate as described in the next project. Lay the back plate on the iPhone with a couple of millimeters overlapping

DOUBLE GEEKERY

Replace the Pentalobe Screws on Your iPhone with Phillips Screws

If your iPhone has pentalobe screws, and you're planning to open it more than once, consider replacing the screws with Phillips screws. The pentalobe screws are much easier to strip than the Phillips screws. If you strip the screws, you'll need to drill them out. This is no fun, even if you're dexterous.

So if your iPhone has pentalobe screws, it's a good idea to spend a couple of dollars on a set of replacement Phillips screws. You can find these on eBay and many other sites by searching using terms such as **iphone 4s back cover phillips screws.**

Whether you decide to replace the screws or not, be very careful when taking the pentalobe screws out so as to avoid stripping them. After taking the iPhone apart and putting it back together, check for wear on the pentalobe screwdriver. Chances are you'll find the screwdriver is worn even after so little use. If so, discard the screwdriver now, because it'll strip the screws next time.

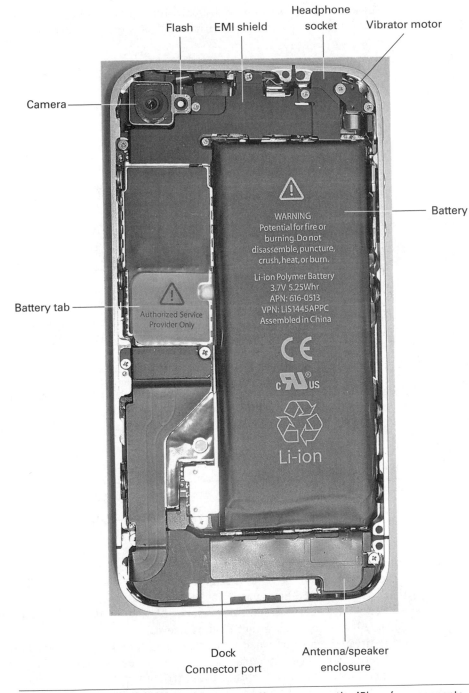

Flash   EMI shield

Headphone socket

Vibrator motor

Camera

Battery

WARNING
Potential for fire or
burning. Do not
disassemble, puncture,
crush, heat, or burn.

Li-ion Polymer Battery
3.7V 5.25Whr
APN: 616-0513
VPN: LIS1445APPC
Assembled in China

Battery tab

Authorized Service
Provider Only

Li-ion

Dock
Connector port

Antenna/speaker
enclosure

FIGURE 6-37 After taking the back plate off, you can see the iPhone's components.

at the top, and then slide it down so that the catches engage. You can then put the original screws back in—or put in replacement screws if you bought them.

Project 48: Put a Custom Back Plate on Your iPhone

If you want your iPhone to stand out from all the others around, you can put a custom back plate on it. This is an easy way to make your iPhone different—and to give it a bit more protection if you want.

You can find plenty of custom back plates on the Internet. Start on eBay and search for **iphone 4s back plate** and see what comes up.

 The iPhone 4S and the iPhone 4 look mighty similar, but their back plates are different. So make sure the back plate you get is for the correct model; otherwise it won't fit.

Once you get the plate, power the iPhone down and take the back plate off as described in the previous project. Put the custom back plate on instead, and you'll have a distinctive iPhone. (Figure 6-38 shows an example.)

FIGURE 6-38   You can put a custom back plate on your iPhone to make it distinctive.

Project 49: Add NFC to Your iPhone

As you've seen in this book, your iPhone has pretty amazing capabilities. But here's one capability it doesn't have that some other smart phones do have—near-field communications capability.

Near-field communications (NFC) is the technology that enables a smart card to communicate with a card reader over the air, so that you don't need to insert the card in the reader. For example, you can keep the smart card inside your wallet, and then just wave the wallet at the card reader—no need to open the wallet.

Or you can keep the smart card in your phone. Various other smart phones, such as Nokia's N9, have NFC capabilities, and pundits have been expecting Apple to add NFC to the iPhone. At this writing, Apple hasn't yet done so—but you can add it yourself.

To add NFC to your iPhone, all you need to do is get a miniature NFC card from your bank. Various banks offer these cards, but few shout about them, so it's well worth asking.

Once you've got the card, take the iPhone's back plate off (as described in Project 47), place the card on top of the battery, and replace the back plate. You'll then be able to make NFC payments using your iPhone.

 If your bank can't provide an NFC card small enough to fit inside your iPhone, put the card in your iPhone's case instead. You'll get a similar effect, even if the execution isn't as neat.

Project 50: Restore Your iPhone to Its Jail

After jailbreaking your iPhone as described at the beginning of this chapter, you may find you need to undo the jailbreak. This project shows you how to do so.

 When you restore your iPhone to its jail, you get rid of Cydia and all the jailbroken apps you've installed.

To restore your iPhone to its jail, follow these steps:

1. Connect your iPhone to your computer, and wait for it to appear in the Source list in iTunes.
2. Click your iPhone's entry in the Source list to display the iPhone screens.
3. Click the Summary tab if it's not already displayed.

4. Click the Restore button. iTunes displays a confirmation dialog box, as shown here, to make sure you know that you're about to erase all the data from the device.

Are you sure you want to restore the iPhone "iPhone" to its factory settings? All of your media and other data will be erased.

After this process is complete, you will have the option to restore your contacts, calendars, text messages and other settings.

Cancel   Restore

If a new version of the iPhone software is available, iTunes prompts you to restore and update the device instead of merely restoring it. Click the Restore And Update button if you want to proceed; otherwise, click the Cancel button.

5. Click the Restore button to close the dialog box. iTunes wipes the device's contents, and then restores the software, showing you its progress while it works.

6. At the end of the restore process, iTunes restarts your iPhone. iTunes displays an information message box for ten seconds while it does so. Either click the OK button or allow the countdown timer to close the message box automatically.

7. After your iPhone restarts, it appears in the Source list in iTunes. Instead of the iPhone's regular tabbed screens, the Set Up Your iPhone screen appears (shown in Figure 6-32, earlier in this chapter).

8. To restore your data, make sure the Restore From The Backup Of option button is selected, and verify that the correct iPhone appears in the drop-down list.

9. Click the Continue button. iTunes restores your data and then restarts the iPhone, displaying another countdown message box while it does so. Either click the OK button or allow the countdown timer to close the message box automatically.

10. After your iPhone appears in the Source list in iTunes following the restart, you can use it as normal.

Index

Symbols and Numbers

S

Satellite button, on Find My iPhone screen, 170, 171
Save Movie Wizard, in Windows Movie Maker, 48–49
Save2PDF, 113
SBSettings
    finding iPhone's IP address, 229–230
    installing, 228
Scan Applications switch, on PKGBackup screen, 223–224
Scan Packages switch, on PKGBackup screen, 223–224
Screen Sharing options, for setting up remote control, 188
screws, opening iPhone and, 259–260
Search screen, in Cydia, 218
Secret area, on Add Configuration screen, 203
Sections screen, in Cydia, 217
Secure Shell (SSH)
    connecting to iPhone via, 231–232
    turning on, 229–230
security, iPhone
    applying restrictions, 151
    Auto-Lock feature, 152
    passcode lock for, 152–157
Security option, for ipCam, 94–95
Sega Genesis, 257, 258
Select Themes button, on WinterBoard screen, 254
self-portraits, high quality
    mounting iPhone on tripod, 72–73
    with rear-facing camera, 71–72
    self-timer function for, 73–74
self-timer function, in camera app, 73–74
Send All Traffic area, on Add Configuration screen, 203
Send Message dialog box, on Find My iPhone screen, 173
Server area, on Add Configuration screen, 202–203
Server Password screen, on Mocha VNC screen, 196
Serversman, 126
Set Network Location dialog box, in Windows, 181–182, 183
Set Passcode screen, 155–156
Set Password dialog box, for iPhone backup, 210–211
Set Up Your iPhone screen, 165, 251–252
Settings button, Air Cam, 92–93
Settings screen, FiLMiC Pro, 76, 77
Settings screen, Gorillacam, 74
Settings screen, iPhone
    connecting to VPN, 204–205
    for Keynote Remote, 148
    setting up Air Sharing, 126, 127
    setting up Home Sharing, 22
    setting up iTunes Match, 31
    setting up Photo Stream, 58–59
Settings screen, PKGBackup, 223, 225
Setup-K-Lite Codec Pack installer, 84–85
Share And Print Screen, in iWork, 111
Share My Library On My Local Network check box, 23–24
Shared category, Home Sharing library in, 20
Shared screen, for playing shared music, 25
sharing iPhone's Internet connection, 179–185
Sharing preferences pane, remote control and, 188
Sharing screen, 119, 126–128
Sharing Security screen, Air Sharing and, 126–128
Sharing tab, of iTunes dialog box, 23, 24
Shortcut screen, 101
shortcuts, text, 100–101, 138
Show button, on Mail, Contacts, Calendars screen, 134
Show More Choices link, in Save Movie Wizard, 49
Signature screen, 137, 138
SIM locking, 175–176
Simple Mode option button, on Setup-K-Lite Codec Pack installer, 84

simple passcode lock
    features of, 153–154
    setting, 155–157
Siri, dictating text to, 101–102
Siri switch, on Passcode Lock screen, 157
Site Manager dialog box, FileZilla, 230–231
Skype
    making calls with, 207–208
    setting up, 206–207
Sleep/Wake button, 152, 163, 166
Snapshot button, Air Cam, 92–93
Software Keyboard Layout box, on Keyboards screen, 106
software reset, 162–163
software, unlocking iPhones using, 178–179
songs. *See* audio
Sony PlayStation, 257, 258
sounding board, connecting iPhone to, 44
Source Selection button, Air Cam, 92–93
Sources screen, in Cydia, 233–234
spam messages, 137
speakers, external, 10–11
special effects, for electric guitar, 36–38
Springboard, restarting, 218, 254–255
Sprint, CDMA iPhones and, 176
SSH. *See* Secure Shell (SSH)
Standard button, on Find My iPhone screen, 170, 171
Start Time value, 26
Startup screen, in Air Cam Options dialog box, 90
Status readout, on VPN screen, 205
Steadicam
    building, 79–82
    mounting iPhone on, 78
    parts needed for, 78–79
    as stabilization device, 77
stereo systems
    connecting iPhone to, 11–12
    using Bluetooth/radio transmitter connection, 13
    using iPhones as car, 16–18
stolen iPhones
    displaying message, 172–173
    locking with passcode, 172
    wiping contents, 173–174
StompBox controller, 37–38
Stop Time value, 26
storage areas for documents, 122
Storage screen, in Cydia, 242–243
Store/Turn Off iTunes Match option, 30
StudioApp, 40–41
subnets, Library Sharing and, 23
Subscribe button, on iTunes Match screen, 29, 30
Successfully Entered DFU Mode! dialog box, in
    PwnageTool, 249–250
Super Nintendo, 257, 258
symbolic links (symlinks), 235
symbols, for Siri dictation, 102
Sync Apps button, 7
syncing iPhone
    with current computer, 3
    to different apps computer, 6–7
    to different information computer, 5–6
    to different iTunes library, 3–4
    to different photos computer, 8–9
    restrictions on, 2
System box, on Storage screen, 242–243
System Preferences, Network Preferences pane in,
    183–184
System Properties dialog box, remote control and, 186–187